Set Theory and Foundations of Mathematics

An Introduction to Mathematical Logic

Second Edition

Volume I
Set Theory

Douglas Cenzer
University of Florida, USA

Christopher Porter
Drake University, USA

Jindřich Zapletal
University of Florida, USA

World Scientific

NEW JERSEY · LONDON · SINGAPORE · BEIJING · SHANGHAI · HONG KONG · TAIPEI · CHENNAI

Published by

World Scientific Publishing Co. Pte. Ltd.
5 Toh Tuck Link, Singapore 596224
USA office: 27 Warren Street, Suite 401-402, Hackensack, NJ 07601
UK office: 57 Shelton Street, Covent Garden, London WC2H 9HE

Library of Congress Control Number: 2024032190

British Library Cataloguing-in-Publication Data
A catalogue record for this book is available from the British Library.

**SET THEORY AND FOUNDATIONS OF MATHEMATICS: AN INTRODUCTION
TO MATHEMATICAL LOGIC**
Volume I: Set Theory
Second Edition

ISBN 978-981-12-9783-0 (hardcover)
ISBN 978-981-12-9784-7 (ebook for institutions)
ISBN 978-981-12-9785-4 (ebook for individuals)

For any available supplementary material, please visit
https://www.worldscientific.com/worldscibooks/10.1142/13974#t=suppl

Desk Editors: Soundararajan Raghuraman/Lai Fun Kwong

Typeset by Stallion Press
Email: enquiries@stallionpress.com

Preface to Second Edition

This book was developed over many years from class notes for a set theory course at the University of Florida. This course has been taught to advanced undergraduates as well as lower level graduate students. The notes have been used more than 30 times as the course has evolved from seminar-style toward a more traditional lecture.

Axiomatic set theory, along with logic, provides the foundation for higher mathematics. This book is focused on the axioms and how they are used to develop the universe of sets, including the integers, rational and real numbers, and transfinite ordinal and cardinal numbers. There is an effort to connect set theory with the mathematics of the real numbers. There are details on various formulations and applications of the Axiom of Choice. Several special topics are covered. The rationals and the reals are studied as dense linear orderings without end points. The possible types of well-ordered subsets of the rationals and reals are examined. The possible cardinality of sets of reals is studied. The Cantor space $2^{\mathbb{N}}$ and Baire space $\mathbb{N}^{\mathbb{N}}$ are presented as topological spaces. Ordinal arithmetic is developed in great detail. The topic of the possible models of fragments of the axioms is examined. As part of the material on the axioms of set theory, we consider models of various subsets of the axioms, as an introduction to consistency and independence. Another interesting topic we cover is an introduction to Ramsey theory.

It is reasonable to cover most of the material in a one-semester course, with selective omissions. Chapter 2 is a review of sets and logic and should be covered as needed in one or two weeks. Chapter 3 introduces the Axioms of Zermelo–Fraenkel, as well as the Axiom of

Choice, in about two weeks. Chapter 4 develops the Natural Numbers, induction and recursion, and introduces cardinality, taking two or three weeks. Chapter 5 on Ordinal Numbers includes transfinite induction and recursion, well orderings, and ordinal arithmetic, in two or three weeks. Chapter 6 covers equivalent versions and applications of the Axiom of Choice, as well as Cardinality, in about two or three weeks. The Real Numbers are developed in Chapter 7, with discussion of dense and complete orders, countable and uncountable sets of reals, and a brief introduction to topological spaces such as the Baire space and Cantor space, again in two or three weeks. If all goes well, this leaves about one week each for the final two chapters: models of set theory and an introduction to Ramsey theory.

This book contains more than 300 exercises which test the students understanding and also enhance the material.

The authors have enjoyed teaching from these notes and are very pleased to share them with a broader audience. We would like to thank the readers of the first edition for many helpful suggestions. A special thanks to James Dudziak for a detailed list of comments.

Douglas Cenzer,
Chris Porter,
Jindra Zapletal
Gainesville, Florida, May 2024

About the Authors

Douglas Cenzer is Professor Emeritus at the University of Florida, where he was Department Chair from 2013 to 2018. He has to his credit more than 120 research publications, specializing in mathematical logic, in particular computability, complexity, and randomness. He has supervised 13 Ph.D. students. Cenzer is an editor for the journal *Archive for Mathematical Logic*. He joined the University of Florida in 1972 after receiving his Ph.D. in mathematics from the University of Michigan. He has held visiting positions at the MSRI, Berkeley, and at the Newton Institute, Cambridge.

Christopher Porter is Associate Professor of Mathematics and Computer Science at Drake University and the Director of Drake's Artificial Intelligence Program. His research interests also include computability theory, algorithmic randomness, and the philosophy of mathematics. Porter received his Ph.D. from the University of Notre Dame in 2012, was an NSF International Fellow at Université Paris 7 from 2012 to 2014, and was a Postdoctoral Associate at the University of Florida from 2014 to 2016, before joining Drake University in 2016.

Jindrich Zapletal is Professor of Mathematics at the University of Florida, specializing in mathematical logic and set theory. He received his Ph.D. in 1995 from the Pennsylvania State University and held postdoctoral positions at MSRI Berkeley, Caltech, and Dartmouth College before joining the University of Florida in 2000.

Contents

Chapter 1

Introduction

Set theory and mathematical logic compose the foundation of pure mathematics. Using the axioms of set theory, we can construct our universe of discourse, beginning with the natural numbers, moving on with sets and functions over the natural numbers, integers, rationals, and real numbers, and eventually developing the transfinite ordinal and cardinal numbers. Mathematical logic provides the language of higher mathematics which allows one to frame the definitions, lemmas, theorems, and conjectures which form the every day work of mathematicians. The axioms and rules of deduction set up the system in which we can prove our conjectures, thus turning them into theorems. Mathematical logic and set theory are also flourishing areas of current mathematical research, such as Ramsey theory. We hope the readers of this book will be inspired to further study in this field.

Chapter 2 begins with elementary naive set theory, including the algebra of sets under union, intersection, and complement and their connection with elementary logic. This chapter introduces the notions of relations, functions, equivalence relations, orderings, and trees. The fundamental notion is *membership*, that is, one set x being a member or element of a second set y; this is written as $x \in y$. Then one set x is a subset of another set y, written as $x \subseteq y$, if every element of x is also an element of y.

Chapter 3 introduces the axioms of Zermelo and Fraenkel. A set should be determined by its elements. Thus, the Axiom of Extensionality states that two sets are equal if and only if they contain exactly the same elements. Some basic axioms provide the existence of simple sets. For example, the Empty Set Axiom asserts the existence of the set $\emptyset$ with no elements. The Axiom of Pairing provides for any two sets x and y a set $\{x, y\}$ with exactly the two members x and y. The Union Axiom provides the union $x \cup y$ of any two given sets, as well as the more general union $\bigcup A$ of a family A of sets. With these we can create sets with three or more elements, for example, $\{a, b, c\} = \{a, b\} \cup \{b, c\}$. The Power Set Axiom collects together into one set $\mathcal{P}(A)$ all subsets of a given set A. The Axiom of Infinity postulates the existence of an infinite set and thus provides for the existence of the set $\mathbb{N}$ of natural numbers. The Axiom of Comprehension provides the existence of the definable subset $\{x \in A : P(x)\}$ of elements of a given set A which satisfy a property P. For example, given the set $\mathbb{N}$ of natural numbers, we can define the set of even numbers as $\{x : (\exists y)x = y + y\}$. The Axiom of Replacement provides the existence of the image $F[A]$ of a given set A under a definable function. The somewhat controversial Axiom of Choice states that for any family $\{A_i : i \in I\}$ of non-empty sets, there is a function F with domain I such that $F(i) \in A_i$ for all $i \in I$. This might seems to be an obvious fact, but it has very strong consequences. In particular, the Axiom of Choice implies the Well-Ordering Principle that every set can be well ordered. A well-ordering $\lhd$ of a set A is an ordering with no descending sequences $a_1 \rhd a_2 \rhd \ldots$. So, the integers can be well ordered by $0 \lhd 1 \lhd -1 \lhd 2 \lhd \ldots$. However, any attempt to well order the set of real numbers will reveal that this is not so obvious after all. Finally, the Axiom of Regularity states that every set A contains a $\in$-minimal element, that is, a set $x \in A$ such that, for all $y \in A$, $y \notin x$. In particular, this implies that no set can belong to itself, and therefore there can be no universal set of all sets. The Axiom of Regularity implies that there is no chain of sets $A_0, A_1, \ldots$ such that $A_{n+1} \in A_n$ for all n. This principle is needed to prove theorems by induction on sets, in the same way that the standard well-ordering on the natural numbers leads to the principle of induction.

Chapter 4 introduces the notion of cardinality, including finite versus infinite, and countable versus uncountable sets. We define the von Neumann natural numbers $N = \{0, 1, 2, \ldots\}$ in the context of set theory. The Induction Principle for natural numbers is established. The methods of recursive and inductive definability over the natural numbers are used to define operations including addition and multiplication on the natural numbers. These methods are also used to define the transitive closure of a set A as the closure of A under the *union* operator and to define the hereditarily finite sets as the closure of 0 under the powerset operator. The Schröder–Bernstein theorem is presented, as well as Cantor's theorem, which shows that the set of subsets of natural numbers is uncountable, and thus the set of reals is also uncountable.

Chapter 5 covers ordinal numbers and their connection with well-orderings. The notions of recursive definitions and the principle of induction on the ordinals are developed. The hierarchy V_α of sets is developed and the notion of *rank* is defined. The standard operations of addition, multiplication, and exponentiation of ordinal arithmetic are defined by transfinite recursion. Various properties of ordinal arithmetic, such as the commutative, associative, and distributive laws, are proved using transfinite induction. This culminates in the Cantor Normal Form Theorem. Well-ordered subsets of the standard real ordering are studied. It is shown that every countable well-ordering is isomorphic to a subset of the rationals and that any well-ordered set of reals is countable.

Chapter 6 is focused on cardinal numbers and the Axiom of Choice. Zorn's lemma and the well-ordering principle are shown to be equivalent to the Axiom of Choice. Zorn's lemma is used to prove the prime ideal theorem and to show that every vector space has a basis. Cardinal numbers are defined and it is shown that, under the Axiom of Choice, every set has a unique cardinality. Hartog's lemma, that every cardinal number has a successor, is proved, thus establishing the existence of uncountable cardinals. The operations of cardinal arithmetic are defined. The Continuum Hypothesis, that the reals have cardinality $\aleph_1$, is formulated. It is shown that the reals cannot have cardinality $\aleph_\omega$. The notion of cofinality and regular cardinals are defined, as well as weakly and strongly inaccessible.

Chapter 7 makes the connection between set theory and the standard mathematical topics of algebra, analysis, and topology. The integers, rationals, and real numbers are constructed inside of the universe of sets, starting from the natural numbers. The rationals are characterized, up to isomorphism, as the unique countable dense linear order without end points. The reals are characterized, up to isomorphism, as the unique complete dense order without end points containing a countable dense subset. The notions of accumulation point and point of condensation are discussed. There is a careful proof of the Cantor–Bendixson theorem, that every closed set of reals can be expressed as a disjoint union of a countable set and a perfect closed set. There is a brief introduction to topological spaces. The Cantor space $2^{\mathbb{N}}$ and Baire space $\mathbb{N}^{\mathbb{N}}$ are studied. It is shown that a subset of $2^{\mathbb{N}}$ is closed if and only if it can be represented as the set of infinite paths through a tree.

Chapter 8 introduces the notion of a *model of set theory*. Conditions are given under which a given set A can satisfy certain of the axioms, such as the union axiom, the Power Set axiom, and so on. It is shown that the hereditarily finite sets satisfy all axioms except for the Axiom of Infinity. The topic of the possible models of fragments of the axioms is examined. In particular, we consider the axioms that are satisfied by V_α when α is, for example, a limit cardinal or an inaccessible cardinal. The hereditarily finite and hereditarily countable, and more generally hereditarily $< \kappa$ sets are also studied in this regard. The hereditarily countable sets are shown to satisfy all axioms except Regularity. This culminates in the proof that V_κ is a model of ZF if and only if κ is a strongly inaccessible cardinal.

Chapter 9 is a brief introduction to Ramsey theory, which studies partitions. This begins with some finite versions of Ramsey's theorem and related results. There is a proof of Ramsey's theorem for the natural numbers as well as the Erdős–Rado theorems, for pairs. Uncountable partitions are also studied.

This additional material gives the instructor options for creating a course which provides the basic elements of set theory and logic, as well as making a solid connection with many other areas of mathematics.

Chapter 2

Review of Sets and Logic

In this chapter, we review some of the basic notions of set theory and logic needed for the rest of this book. There is a very close connection between the Boolean algebra of sets and the formulas of predicate logic. We present some aspects of so-called naive set theory and indicate the methods of proof used there as a foundation for more advanced notions and theorems. Topics here include functions and relations, in particular, orderings and equivalence relations, presented at an informal level. We return to these topics in a more formal way once we begin to study the axiomatic foundation of set theory. Students who have had a transition course to higher mathematics, such as a course in sets and logic, should be able to go right to the following chapter.

2.1   The Algebra of Sets

In naive set theory, there is a *universe* U of all elements. For example, this may be the set $\mathbb{R}$ of real numbers or the set $\mathbb{N} = \{0, 1, 2, \dots\}$ of natural numbers, or perhaps some finite set. The fundamental relation of set theory is that of membership. For a subset A of U and an element a of A, we write $a \in A$ to mean that a belongs to A or is an element of A. The family $\mathcal{P}(U)$ of subsets of U has the natural Boolean operations of union, intersection, and complement, as follows.

Definition 2.1.1. For any element a of U and any subsets A and B of U,

1. $a \in A \cup B$ if and only if $a \in A \ \vee \ a \in B$;
2. $a \in A \cap B$ if and only if $a \in A \ \wedge \ a \in B$;
3. $a \in A^C$ if and only if $\neg \, a \in A$.

Here we use the symbols $\vee$, $\wedge$, and $\neg$ to denote the logical connectives *or*, *and*, and *not*. We frequently write $x \notin A$ as an abbreviation for $\neg \, x \in A$.

The convention is that two sets A and B are equal if they contain the same elements. That is,

$$A = B \iff (\forall x)[x \in A \iff x \in B].$$

This is codified in the Axiom of Extensionality, one of the axioms of Zermelo Fraenkel set theory which is presented in detail in Chapter 3. The family of subsets of U compose a Boolean algebra, that is, they satisfy certain properties, such as the associative, commutative, and distributive laws. We consider some of these now and leave others to the exercises. We put in all of the details at first and later on leave some of them to the reader.

Proposition 2.1.2 (Commutative Laws). *For any sets A and B,*

1. $A \cup B = B \cup A$;
2. $A \cap B = B \cap A$.

Proof.   (1) Let x be an arbitrary element of U. We want to show that, for any $x \in U$, $x \in A \cup B \iff x \in B \cup A$. By propositional logic, this means we need to show that $x \in A \cup B \to x \in B \cup A$ and that $x \in B \cup A \to x \in A \cup B$. To prove the first implication, we need to suppose that $x \in A \cup B$ and then deduce that $x \in B \cup A$. We now proceed as follows. Suppose that $x \in A \cup B$. Then by Definition 2.1.1, $x \in A$ or $x \in B$. It follows by propositional logic that $x \in B \vee x \in A$. Hence, by Definition 2.1.1, $x \in B \cup A$. Thus, we have shown $x \in A \cup B \to x \in B \cup A$. A similar argument shows that $x \in B \cup A \to x \in A \cup B$. Then, $x \in A \cup B \iff x \in B \cup A$. Since x was arbitrary, we have $(\forall x)[x \in A \cup B \iff x \in B \cup A]$. It then follows by Extensionality that $A \cup B = B \cup A$.

Part (2) is left to the exercises. $\square$

The notion of subset, or inclusion, is fundamental.

Definition 2.1.3. For any sets A and B,

1. $A \subseteq B \iff (\forall x)[x \in A \to x \in B]$. We say that A is *included* in B if $A \subseteq B$.
2. $A \subsetneq B \iff A \subseteq B \land A \neq B$.

Proposition 2.1.4 (Associative Laws).

1. $A \cap (B \cap C) = (A \cap B) \cap C$;
2. $A \cup (B \cup C) = (A \cup B) \cup C$.

Proof. (1) $A \cap (B \cap C) = (A \cap B) \cap C$. Let x be an arbitrary element of U and suppose that $x \in A \cap (B \cap C)$. Then, by Definition 2.1.1, we have $x \in A$ and $x \in B \cap C$ and therefore $x \in B$ and $x \in C$. It follows by propositional logic that $x \in A \land x \in B$, and thus $x \in A \cap B$. Then, by propositional logic, $(x \in A \cap B) \land x \in C$. Thus, by Definition 2.1.1, $x \in (A \cap B) \cap C$. Thus, $x \in A \cap (B \cap C) \to x \in (A \cap B) \cap C$. A similar argument shows that $x \in (A \cap B) \cap C \to x \in (A \cap (B \cap C)$. Since x was arbitrary, we have $(\forall x)[x \in A \cap (B \cap C) \to x \in (A \cap B) \cap C]$. It now follows by Extensionality that $A \cap (B \cap C) = (A \cap B) \cap C$.

Part (2) is left to the exercises. □

The following proposition can help simplify a proof that two sets are equal.

Proposition 2.1.5. *For any sets A and B, $A = B \iff A \subseteq B \land B \subseteq A$.*

Proof. Suppose first that $A = B$. This means that $(\forall x)[x \in A \iff x \in B]$. Now, let $x \in U$ be arbitrary. Then, $x \in A \iff x \in B$. It follows from propositional logic that $x \in A \to x \in B$ and also $x \in B \to x \in A$. Since x was arbitrary, we have $(\forall x)[x \in A \to x \in B]$ and $(\forall x)[x \in B \to x \in A]$. Then, by Definition 2.1.3, it follows that $A \subseteq B$ and $B \subseteq A$.

Next, suppose that $A \subseteq B$ and $B \subseteq A$. The steps above can be reversed to deduce that $A = B$. □

The empty set $\emptyset$ is defined by the following property:

$$(\forall x) \ x \notin \emptyset.$$

It is easy to see that $\emptyset = U^C$ and that $\emptyset^C = U$. This is left as an exercise.

Proposition 2.1.6 (DeMorgan's Laws). *For any sets A and B,*

1. $(A \cup B)^C = A^C \cap B^C$;
2. $(A \cap B)^C = A^C \cup B^C$.

Proof. (1) We prove this by a sequence of equivalent statements. Let $x \in U$ be arbitrary. Then, $x \in (A \cup B)^C$ if and only if $x \notin A \cup B$ if and only if $\neg(x \in A \vee x \in B)$ if and only if $x \notin A \wedge x \notin B$ if and only if $x \in A^C \wedge x \in B^C$ if and only if $x \in A^C \cap B^C$.

Part (2) is left to the exercises. $\qquad\square$

The universal set U and the empty set are the identities of the Boolean algebra $\mathcal{P}(U)$. This is spelled out in the following proposition.

Proposition 2.1.7 (Identity Laws). *For any set A,*

1. $A \cup A^C = U$;
2. $A \cap A^C = \emptyset$.

Proof. (1) One inclusion follows from the fact that $B \subseteq U$ for all sets B. For the other inclusion, let $x \in U$ be arbitrary. It follows from propositional logic (the so-called law of excluded middle) that $x \in A \vee \neg x \in A$. Then, by Definition 2.1.1, $x \in A \vee x \in A^C$ and then $x \in A \cup A^C$. Thus, $U \subseteq A \cup A^C$.

(2) This follows from (1) using DeMorgan's laws. Given part (1) that $A \cup A^C = U$, we obtain $\emptyset = U^C = (A \cup A^C)^C = A^C \cap (A^C)^C = A^C \cap A = A \cap A^C$. $\qquad\square$

The inclusion relation may be seen to be a partial ordering. We have just seen above that it is antisymmetric, that is, $A \subseteq B$ and $B \subseteq A$ imply $A = B$. Certainly, this relation is reflexive, that is, $A \subseteq A$. Transitivity is left to the exercises.

Inclusion may be defined from the Boolean operations in several ways.

Proposition 2.1.8. *The following are equivalent:*

1. $A \subseteq B$;
2. $A \cap B = A$;
3. $A \cup B = B$.

Proof. We show that (1) and (2) are equivalent and leave the other equivalence to the exercises:

(1) $\implies$ (2): Assume that $A \subseteq B$. Let x be arbitrary. Then, $x \in A \to x \in B$. Now, suppose that $x \in A$. Then, $x \in B$ and hence $x \in A \wedge x \in B$ so that $x \in A \cap B$. Thus, $A \subseteq A \cap B$. Next, suppose that $x \in A \cap B$. Then, $x \in A \wedge x \in B$, so certainly $x \in A$. Thus, $A \cap B \subseteq A$. It follows that $A \cap B = A$, as desired.

(2) $\implies$ (1): Suppose that $A \cap B = A$. Let x be arbitrary and suppose that $x \in A$. Since $A \cap B = A$, it follows that $x \in A \cap B$. That is, $x \in A \wedge x \in B$ so that $x \in B$. Hence, $A \subseteq B$. $\square$

We sometimes write $A \setminus B$ for $A \cap B^{\mathsf{C}}$. The proof of the following is left as an exercise.

Proposition 2.1.9. *The following are equivalent:*

1. $A \subseteq B$;
2. $B^{\mathsf{C}} \subseteq A^{\mathsf{C}}$;
3. $A \setminus B = \emptyset$.

There are some interactions between the inclusion relation and the Boolean operations, in the same way that inequality for numbers interacts with the addition and multiplication operations.

Proposition 2.1.10. *For any sets A, B, and C,*

1. *if $B \subseteq A$ and $C \subseteq A$, then $B \cup C \subseteq A$;*
2. *if $A \subseteq B$ and $A \subseteq C$, then $A \subseteq B \cap C$.*

Proof. (1) Assume that $B \subseteq A$ and $C \subseteq A$. Let x be arbitrary and suppose that $x \in B \cup C$. This means that $x \in B \vee x \in C$. There are two cases. Suppose first that $x \in B$. Since $B \subseteq A$, it follows that $x \in A$. Suppose next that $x \in C$. Since $C \subseteq A$, it follows again that $x \in A$. Hence, $x \in B \cup C \to x \in A$. Since x was arbitrary, we have $B \cup C \subseteq A$, as desired.

The proof of Part (2) is left to the exercises. $\square$

Exercises for Section 2.1

Exercise 2.1.1. Prove the Commutative Law for intersection, that is, for any sets A and B, $A \cap B = B \cap A$.

Exercise 2.1.2. Prove the Associative Law for union, that is, for any sets A, B, and C, $A \cup (B \cup C) = (A \cup B) \cup C$.

Exercise 2.1.3. Prove the Distributive Laws, that is, for any sets A, B, and C, $A \cup (B \cap C) = (A \cup B) \cap (A \cup C)$ and $A \cap (B \cup C) = (A \cap B) \cup (A \cap C)$.

Exercise 2.1.4. Show that for any set A, $\emptyset \subseteq A$ and $A \subseteq U$.

Exercise 2.1.5. Show that for any set A, $(A^{\complement})^{\complement} = A$.

Exercise 2.1.6. Show that for any set A, $A \cup \emptyset = A$ and $A \cap U = A$.

Exercise 2.1.7. Show that for any set A, $A \cap \emptyset = \emptyset$ and $A \cup U = U$.

Exercise 2.1.8. Complete the argument that the relation $\subseteq$ is a partial ordering by showing that it is transitive, that is, if $A \subseteq B$ and $B \subseteq C$, then $A \subseteq C$.

Exercise 2.1.9. Show that for any sets A and B, $A \subseteq B$ if and only if $A \cup B = B$.

Exercise 2.1.10. Show that the following are equivalent:

1. $A \subseteq B$;
2. $B^{\complement} \subseteq A^{\complement}$;
3. $A \setminus B = \emptyset$.

Exercise 2.1.11. Show that or any sets A, B, and C, $A \subseteq B$ & $A \subseteq C$ implies that $A \subseteq B \cap C$.

2.2 Relations

Relations play a fundamental role in mathematics. Of particular interest are orderings, equivalence relations, and graphs. The notion of a graph is quite general. That is, a *graph* $G = (V, E)$ is simply a set V of *vertices* and a binary relation E on V. In a *directed* graph, a pair $(u, v) \in E$ is said to be an *edge* from u to v. A graph (V, E) is said to be *undirected* if $(u, v) \in E$ implies $(v, u) \in E$ for all $u, v \in V$.

A key notion here is that of an *ordered pair*. Given two elements a_1, a_2 from our universe U, the ordered pair (a_1, a_2) is defined so that for any two pairs of elements (a_1, b_1) and (b_1, b_2), $(a_1, a_2) = (b_1, b_2) \iff a_1 = b_1$ and $a_2 = b_2$. This is defined carefully in the following chapter, along with the notion of an n-tuple $(a_1, \ldots, a_n)$ of elements.

Definition 2.2.1. Let $A_1, \ldots, A_n$ be sets:

1. The product $A_1 \times A_2 \times \cdots \times A_n = \{(a_1, \ldots, a_n) : \text{each } a_i \in A_i\}$.
2. $A^n = \{(a_1, \ldots, a_n) : \text{each } a_i \in A\}$.

Definition 2.2.2. Let A and B be sets:

1. The product of A and B is defined to be

$$A \times B = \{(a, b) : a \in A \wedge b \in B\}.$$

2. A subset R of $A \times B$ is called a *relation*, specifically a *binary relation*. We sometimes write aRb for $(a, b) \in R$.
3. A subset R of $A_1 \times A_2 \times \cdots \times A_n$ is said to be an n-ary relation.

Proposition 2.2.3. *For any sets A, B, and C,*

1. $A \times (B \cup C) = (A \times B) \cup (A \times C)$ *and* $(B \cup C) \times A = (B \times A) \cup (C \times A)$;
2. $A \times (B \cap C) = (A \times B) \cap (A \times C)$ *and* $(B \cap C) \times A = (B \times A) \cap (C \times A)$;
3. $A \times (B \backslash C) = (A \times B) \backslash (A \times C)$ *and* $(A \backslash B) \times C = (A \times C) \backslash (B \times C)$.

Proof. (1) $(x, y) \in A \times (B \cup C)$ if and only if $x \in A \wedge y \in B \cup C$, if and only if $x \in A \wedge (y \in B \vee y \in C)$, if and only if $(x \in A \wedge y \in B) \vee (x \in A \wedge y \in C)$, if and only if $(x, y) \in A \times B \vee (x, y) \in A \times C$, if and only if $(x, y) \in (A \times B) \cup (A \times C)$. The proof of the second statement in (1) is similar.

The proofs of Parts (2) and (3) are left to the exercises. □

Here are some important examples of binary relations that we return to frequently in what follows.

Example 2.2.4. The standard ordering $x \leq y$ (as well as the strict order $<$) on the real numbers is a binary relation, which also applies to the rational numbers, integers, and natural numbers.

Example 2.2.5. The subset, or inclusion, relation $A \subseteq B$ on sets, read "A is a subset of B", or "A is included in B", is a binary relation.

Example 2.2.6. Let the graph G have vertices (i, j) for integers i, j; this is the lattice of integer points in the plane. Let there be horizontal and vertical edges between adjacent vertices. That is, each (i, j) has four edges, going to $(i-1,j)$, $(i+1,j)$, $(i,j-1)$, and $(i,j+1)$ which means that, for example, $(1,2)E(1,3)$.

Example 2.2.7. The fundamental relation of axiomatic set theory is membership, that is, the relation $x \in y$, for sets x and y. Note that when we study higher set theory, we do not distinguish between points and sets.

Example 2.2.8. For any set A, let $I_A = \{(a,a) \in A \times A : a \in A\}$ be the *identity* relation on A.

Example 2.2.9. The divisibility relation on the set $\mathbb{Z}$ of integers is defined by $x \mid y \iff (\exists z)\, y = xz$.

Here are some useful concepts associated with relations.

Definition 2.2.10. For any sets A and B, and any relation $R \subseteq A \times B$:

1. The inverse R^{-1} of R is $R^{-1} = \{(u,v) : vRu\}$.
2. The domain of R is $Dmn(R) = \{x : (\exists y)\, xRy\}$ and for any set $D \subseteq B$, the inverse image of D is $R^{-1}[D] = \{x \in A : (\exists y \in D)\, xRy\}$.
3. The range of R is $Rng(R) = \{y : (\exists x)\, xRy\}$, and for any $C \subseteq A$, the image $R[C] = \{y \in B : (\exists x \in C)\, xRy\}$.

For example, if xRy is the ordering $x \leq y$, then $xR^{-1}y$ is the ordering $x \geq y$. The inverse of the identity relation is again the identity. On the natural numbers, the domain of strict inequality is $\mathbb{N}$ but the range is just $\mathbb{N}^+$. For the strict ordering on the real numbers, let $A = \{a\}$ be the set with a single element a. Then, $R[A] = (a, \infty)$ and $R^{-1}[A] = (-\infty, a)$. If R is the subset relation $\subseteq$ on U and A is any subset of U, then $R^{-1}[A] = \mathcal{P}(A)$.

Proposition 2.2.11. *For any relations R and S,*

1. $(R \cup S)^{-1} = R^{-1} \cup S^{-1}$;
2. $(R \cap S)^{-1} = R^{-1} \cap S^{-1}$.

Proof. (1) $(x,y) \in (R \cup S)^{-1}$ if and only if $(y,x) \in R \cup S$ if and only if $(y,x) \in R \lor (y,x) \in S$ if and only if $(x,y) \in R^{-1} \lor (x,y) \in S^{-1}$ if and only if $(x,y) \in R^{-1} \cup S^{-1}$.

Part (2) is left to the exercise. $\qquad\square$

Proposition 2.2.12. *For any relations R and S,*

1. $Dmn(R \cap S) \subseteq Dmn(R) \cap Dmn(S)$;
2. $Rng(R \cap S) \subseteq Rng(R) \cap Rng(S)$.

Proof. (1) Suppose $x \in Dmn(R \cap S)$. Then, for some y, $(x, y) \in R \cap S$. Choose some b so that $(x, b) \in R \cap S$. Then, $(x, b) \in R$ and $(x, b) \in S$. It follows that $(\exists y)(x, y) \in R$ and $(\exists y)(x, y) \in S$. Thus, $x \in Dmn(R) \wedge x \in Dmn(S)$, and therefore $x \in Dmn(R) \cap Dmn(S)$. The above steps can be reversed to obtain the other inclusion. The proof of (2) is left to the reader. □

Definition 2.2.13. If $R \subseteq B \times C$ and $S \subseteq A \times B$ are relations, then the *composition* $R \circ S$ is defined by $R \circ S = \{(u, v) : (\exists w)(uSw \wedge wRv)\}$.

Example 2.2.14. Here are some illustrations from the examples above:

1. From Example 2.2.4, let R be the strict ordering $x < y$ on the integers. Then, $(x, y) \in R \circ R$ if and only if $y - x \geq 2$.
2. From Example 2.2.6, the graph G of the lattice of integer points in the plane, two points are related in $E \circ E$ if there is a path of length two connecting them.
3. From Example 2.2.8, the identity relation I_A acts like an identity for $\circ$ in that $I_A \circ R = R = R \circ I_A$. The proof is left as an exercise.

Example 2.2.15. For any set A, the set of permutations of A forms a group under composition. This is demonstrated by some of the properties of the following composition.

Here are some important properties of composition.

Proposition 2.2.16. *If $R \subseteq B \times C$ and $S \subseteq A \times B$ are relations, then*

1. $Dmn(R \circ S) \subseteq Dmn(S)$ *and*
2. $Rng(R \circ S) \subseteq Rng(R)$.

Proof. (1) Suppose that $x \in Dmn(R \circ S)$. Then, for some y, $(x, y) \in R \circ S$. This means that there is some v such that $(x, v) \in S$ and $(v, y) \in R$. By the first part, we have $x \in Dom(S)$.
   Part (2) is left as an exercise. □

Proposition 2.2.17. *For any relations R, S, and T,*

1. $(R \cap S) \circ T \subseteq (R \circ T) \cap (S \circ T)$;
2. $R \circ (S \cap T) \subseteq (R \circ S) \cap (R \circ T)$.

Proof. (1) Suppose that $(x, y) \in (R \cap S) \circ T$. Then, for some z, $(x, z) \in T$ and $(z, y) \in R \cap S$. Then, $(z, y) \in R$ and $(z, y) \in S$ so that $(x, y) \in R \circ T \wedge (x, y) \in S \circ T$. Thus, $(x, y) \in (R \circ T) \cap (S \circ T)$.
Part (2) is left as an exercise. □

The following example shows that inequality does not always hold in (1) above.

Example 2.2.18. Define relations R, S, and T on the natural numbers as follows: $T = \{(x, 2x), (x, 3x) : x \in \mathbb{N}\}$, $R = \{(2x, x) : x \in \mathbb{N}\}$, and $S = \{3x, x) : x \in \mathbb{N}\}$. Then, $R \circ T = S \circ T = \{(x, x) : x \in \mathbb{N}\}$ so that $(R \circ T) \cap (S \circ T) = \{(x, x) : x \in \mathbb{N}\}$. On the other hand, $R \cap S = \{(0, 0\}$ so that $(R \cap S) \circ T = \{(0, 0\}$. This also provides an example that equality does not hold in Proposition 2.2.12. That is, $Dmn(R \cap S) = \{0\}$, whereas $Dmn(R) \cap Dmn(S) = \{0, 2, 4 \ldots \} \cap \{0, 3, 6, \ldots \} = \{0, 6, 12, \ldots \}$. Moreover, $Rng(R \cap S) = \{0\}$, but $Rng(R) \cap Rng(S) = \mathbb{N} \cap \mathbb{N} = \mathbb{N}$.

The next proposition says that composition is associative.

Proposition 2.2.19. *If $R \subseteq C \times D$, $S \subseteq B \times C$, and $T \subseteq A \times B$ are relations, then $R \circ (S \circ T) = (R \circ S) \circ T$.*

Proof. ($\subseteq$): Suppose that $(x, y) \in R \circ (S \circ T)$. Then for some $z \in C$, $(x, z) \in S \circ T$ and $(z, y) \in R$. The first statement implies that for some $v \in B$, $(x, v) \in T$ and $(v, z) \in S$. Since $(v, z) \in S$ and $(z, y) \in R$, it follows that $(v, y) \in R \circ S$. Since $(x, v) \in T$, it follows that $(x, y) \in (R \circ S) \circ T$.
The steps above can be reversed to obtain the other inclusion. □

Proposition 2.2.20. *If $R \subseteq B \times C$ and $S \subseteq A \times B$ are relations, then $(R \circ S)^{-1} = S^{-1} \circ R^{-1}$.*

Proof. ($\subseteq$): Suppose $(x, y) \in (R \circ S)^{-1}$. Then, $(y, x) \in R \circ S$. This means that for some $z \in B$, $(y, z) \in S$ and $(z, x) \in R$. It follows that $(z, y) \in S^{-1}$ and $(x, z) \in R^{-1}$. This implies that $(x, y) \in S^{-1} \circ R^{-1}$.
($\supseteq$): Suppose that $(x, y) \in S^{-1} \circ R^{-1}$. Then for some $z \in B$, $(z, y) \in S^{-1}$ and $(x, z) \in R^{-1}$. Thus, $(y, z) \in S$ and $(z, x) \in R$. It follows that $(y, x) \in R \circ S$. This implies that $(x, y) \in (R \circ S)^{-1}$. □

For our discussion of equivalence relations and orderings, we need the following basic concepts about relations.

Definition 2.2.21. For any binary relation R on a set A,

1. R is *reflexive* if for any $x \in A$, xRx,
2. R is *irreflexive* if for any $x \in A$, $\neg xRx$,
3. R is *transitive* if for any $x, y, z \in A$, if both xRy and yRz, then xRz,
4. R is *symmetric* if for any $x, y \in A$, xRy if and only if yRx,
5. R is *antisymmetric* if for any $x, y \in A$, if both xRy and yRx, then $x = y$.

Example 2.2.22. Returning to the examples above, we have the following:

1. From Example 2.2.4, the standard ordering $x \leq y$ on the real numbers is reflexive, transitive, and antisymmetric. The strict order $<$ is irreflexive, transitive, and antisymmetric (in fact, it is never true that $x < y$ and $y < x$).
2. From Example 2.2.5, the subset relation $A \subseteq B$ is reflexive, transitive, and antisymmetric. The last is the property of extensionality. That is, if $A \subseteq B$ and $B \subseteq A$, then A and B contain the same elements and are therefore equal.
3. From Example 2.2.6, the graph G representing the lattice of integer points in the plane is irreflexive, not transitive, but it is symmetric.
4. From Example 2.2.7, the membership relation $x \in y$ is irreflexive, not transitive, and antisymmetric, that is, we can never have $x \in y$ and $y \in x$. This is explained carefully in the following chapter.
5. From Example 2.2.8, the identity relation I_A on a set A is reflexive, transitive, and symmetric. The last two properties follows from the fact that $(x, y) \in I_A$ implies $x = y$.
6. From Example 2.2.9, the divisibility relation on $\mathbb{Z}$ is reflexive and transitive. On the natural numbers, it is also antisymmetric.

Exercises for Section 2.2

Exercise 2.2.1. Show that for any set A, $A \times \emptyset = \emptyset \times A = \emptyset$.

Exercise 2.2.2. Show that for any non-empty sets A and B, $A \times B$ is non-empty.

Exercise 2.2.3. Show that $(A \times B)^{-1} = B \times A$.

Exercise 2.2.4. Show that for any relations R and S, $(R \cap S)^{-1} = R^{-1} \cap S^{-1}$.

Exercise 2.2.5. Show that for any relation R, $(R^{-1})^{-1} = R$.

Exercise 2.2.6. Show that for any sets A, B, and C, $A \times (B \cap C) = (A \times B) \cap (A \times C)$ and $(B \cap C) \times A = (B \times A) \cap (C \times A)$.

Exercise 2.2.7. Prove the following: For any sets A, B, and C,

(a) $A \times (B \setminus C) = (A \times B) \setminus (A \times C)$ and
(b) $(A \setminus B) \times C = (A \times C) \setminus (B \times C)$.

Exercise 2.2.8. Let $a < b$ be real numbers. Find the image and inverse image of $[a, b]$ and (a, b) under $\leq$ and under $<$.

Exercise 2.2.9. For any relation R, show that $Dmn(R^{-1}) = Rng(R)$ and $Rng(R^{-1}) = Dmn(R)$.

Exercise 2.2.10. For any relations R and S, show that $Dmn(R \cup S) = Dmn(R) \cup Dmn(S)$ and $Rng(R \cup S) = Rng(R) \cup Rng(S)$.

Exercise 2.2.11. Let R and S be relations. $Rng(R \cap S) \subseteq Rng(R) \cap Rng(S)$.

Exercise 2.2.12. For any relations R and S, show that

(a) $Dmn(R) \setminus Dmn(S) \subseteq Dmn(R \setminus S)$ and
(b) $Rng(R) \setminus Rng(S) \subseteq Rng(R \setminus S)$.

Exercise 2.2.13. Prove the following:

(a) If $B \cap C = \emptyset$, then $(C \times D) \circ (A \times B) = \emptyset$.
(b) If $B \cap C \neq \emptyset$, then $(C \times D) \circ (A \times B) = A \times D$.

Exercise 2.2.14. Show that if $R \subset A \times A$, then $I_A \circ R = R \circ I_A = R$.

Exercise 2.2.15. Show that for any relations R and S, $Rng(R \circ S) \subseteq Rng(R)$.

Exercise 2.2.16. For any relations R, S, and T, $R \circ (S \cap T) \subseteq (R \circ S) \cap (R \circ T)$.

Exercise 2.2.17. For any relation $R \subseteq A \times A$, if $Dmn(R) = A = Rng(R)$, then $I_A \subseteq R \circ R^{-1}$ and $I_A \subseteq R^{-1} \circ R$.

Exercise 2.2.18. For any relations R, S, and T,
$$(R \cup S) \circ T = (R \circ T) \cup (S \circ T)$$
and $R \circ (S \cup T) = (R \circ S) \cup (R \circ T)$.

Exercise 2.2.19. Prove that for any relations, R, S, and T, $(R \circ S) \cap T$ is empty if and only if $(R^{-1} \circ T) \cap S$ is empty.

Exercise 2.2.20. Let R be a relation and let A, B be arbitrary subsets $Dmn(R)$:

(a) Show that $R[A \cup B] = R[A] \cup R[B]$.
(b) Show that $R[A \cap B] \subseteq R[A] \cap R[B]$.
(c) Show that equality does not always hold in part (b).

Exercise 2.2.21. Show that for any relations R and S and any $A \subseteq Dmn(S)$, $(R \circ S)[A] = R[S[A]]$.

Exercise 2.2.22. Let R and S be relations:

(a) Show that $Dmn(R \circ S) = S^{-1}[Dmn(R)]$.
(b) Show that $Rng(R \circ S) = R[Rng(S)]$.

Exercise 2.2.23. Suppose R is a relation on U. Prove the following:

(a) R is reflexive if and only if $I_U \subseteq R$.
(b) R is irreflexive if and only if $I_U \cap R = \emptyset$.
(c) R is transitive if and only if $R \circ R \subseteq R$.
(d) R is symmetric if and only if $R = R^{-1}$.
(e) R is antisymmetric if and only if $R \cap R^{-1} \subseteq I_U$.
(f) If R is transitive and reflexive, then $R \circ R = R$.

2.3   Functions

Functions are of fundamental importance in mathematics. The integers come equipped with binary addition and multiplication functions. In college algebra and trigonometry, we learn about the exponential function and the sine, cosine, and tangent functions on real numbers. Just as relations may be viewed as sets, functions may be viewed as relations and hence also as sets.

Definition 2.3.1. A relation F on $A \times B$ is a *function* if, for every $x \in Dmn(F)$, there is a unique $y \in Rng(F)$ such that xFy. We write $y = F(x)$ for xFy. If $Dmn(F) = A$ and $Rng(F) \subseteq B$, we say that F *maps* A into B, written as $F : A \to B$. F is *one-to-one*, or *injective*, if F^{-1} is also function. F maps A *onto* B, or is *surjective*, if $Rng(F) = B$. F is *bijective*, or is a *set isomorphism* from A to B, if F is injective and surjective.

Definition 2.3.2. For any sets A and B, B^A is the set of functions mapping A into B.

A function F is said to be *binary*, or in general n-ary, if $Dmn(F) \subseteq A \times A$ (in general A^n) for some set A. Most commonly studied functions are either 1-ary (unary) or binary.

In the calculus, we studied how to determine whether functions were one-to-one and how to find their domain and range. For example, the function $f(x) = x^3$ is both injective and surjective. The exponential function $f(x) = e^x$ is injective but not surjective. The function $f(x) = x^3 - x$ is surjective, but it is not injective, since $f(0) = f(1) = 0$.

In any group G, the function mapping x to its inverse x^{-1} is a set isomorphism.

Equality of functions may be characterized as follows.

Proposition 2.3.3. *Let F and G be two functions mapping set A to set B. Then $F = G$ if and only if $F(x) = G(x)$ for all $x \in A$.*

Proof. Suppose first that $F = G$ and let $x \in A$. Since F and G are functions, there are unique elements b and c of B such that $F(x) = a$ and $G(x) = c$. Then $(x, a) \in F$ and $(x, c) \in G$. Since $F = G$, it follows that both (x, a) and (x, c) are in F. Since F is a function, it follows that $b = c$ so that $F(x) = G(x)$.

Suppose next that $F(x) = G(x)$ for all $x \in A$. Then, for any $a \in A$ and $b \in B$, we have $(a, b) \in F$ if and only if $F(a) = b$, if and only if $G(a) = b$, if and only if $(a, b) \in G$. Thus, $F = G$. $\square$

All of the results about relations also apply to functions, but there are some additional nice properties of functions. Note that for a function $F : A \to B$ and $C \subseteq B$, the inverse image of C under F, $F^{-1}[C]$, is defined by taking the inverse of F as a relation, i.e.,

$$F^{-1}[C] = \{x \in A : F(x) \in C\}.$$

Proposition 2.3.4. *For any function $F : C \to D$ and any subsets A, B of D,*

1. $F^{-1}[A \cap B] = F^{-1}[A] \cap F^{-1}[B]$;
2. $F^{-1}[A \setminus B] = F^{-1}[A] \setminus F^{-1}[B]$.

Proof.   Let $x \in C$. Then, $x \in F^{-1}[A \cap B]$ if and only if $F(x) \in A \cap B$, if and only if $F(x) \in A \wedge F(x) \in B$, if and only if $x \in F^{-1}[A] \wedge x \in F^{-1}[B]$, if and only if $x \in F^{-1}[A] \cap F^{-1}[B]$.

Part (2) is left to the reader. □

It is not hard to see that $F \circ G$ is a function if F and G are functions (see the exercises.) Here are some interesting results about the composition of functions.

Proposition 2.3.5. *Let $F : A \to B$ be a function:*

1. *$F : A \to B$ is one-to-one if and only if, for all C and all $G : C \to A$ and $H : C \to A$, $F \circ G = F \circ H$ implies $G = H$.*
2. *$F : A \to B$ is onto if and only if, for all C and all $G : B \to C$ and $H : B \to C$, $G \circ F = H \circ F$ implies $G = H$.*

Proof.   (1) Let $F : A \to B$. Suppose that F is one-to-one and let $G : C \to A$ and $H : C \to A$. Suppose also that $F \circ G = F \circ H$. Then for any $x \in C$, $F(G(x)) = F(H(x))$. Since F is one-to-one, this implies that $G(x) = H(x)$. Thus, $G = H$.

Next, suppose that $F : A \to B$ is *not* one-to-one and choose $a_1 \neq a_2 \in A$ and $b \in B$ such that $F(a_1) = F(a_2) = b$. Let $C = \{c\}$ and define $G(c) = a_1$ and $H(c) = a_2$ so that $F \circ G(c) = b = F \circ H(c)$ and hence $F \circ G = F \circ H$. However, $G \neq H$.

Part (2) is left as an exercise. □

Next, we consider indexed families of sets. This is an important topic later in connection with the Axiom of Choice. An indexed family $\{A_i : i \in I\}$ of sets may be viewed as a function from I to $\bigcup_{i \in I} A_i$.

Definition 2.3.6. Suppose $\mathcal{A} = \{A_i : i \in I\}$ is an indexed family of sets:

1. The union of this family is $\bigcup_{i \in I} A_i := \{u : (\exists i \in I)\, u \in A_i\}$.
2. The intersection of this family is $\bigcap_{i \in I} A_i := \{u : (\forall i \in I)\, u \in A_i\}$.
3. The Cartesian product of this family is $\prod_{i \in I} A_i = \{f : I \to \bigcup_{i \in I} A_i : (\forall i)[f(i) \in A_i]\}$.
4. For each $i \in I$, the *i-th projection function*, $p_i : \prod_{i \in I} A_i \to A_i$, is defined by $p_i(f) = f(i)$ for all $f \in \prod_{i \in I} A_i$.

Example 2.3.7.

1. Let I be the set of prime numbers and let $A_p = \{n \in \mathbb{N} : p \mid n\}$. Then, $\bigcup_{p \in I} A_p = \{n \in \mathbb{N} : n \neq 1\}$ and $\bigcap_{p \in I} A_p = \{0\}$.
2. Let $I = \mathbb{N}^+ = \{1, 2, \dots\}$ and let $A_n = (-1/n, 1/n)$. Then, $\bigcup_{n \in I} A_n = (-1, 1)$ and $\bigcap_{n \in I} A_n = \{0\}$.
3. Let $I = \mathbb{N}$ and let $A_i = \mathbb{R}$ for all $i \in \mathbb{N}$. Then, $\prod_{i \in \mathbb{N}} A_i$ is the set of infinite sequences of real numbers, which plays a very important role in the study of calculus.
4. The Cantor space is defined to be $\prod_{i \in \mathbb{N}} \{0, 1\}$, the set of infinite sequences of 0's and 1's. This is one of the fundamental spaces of topology.

Now, we can examine unions and intersections of infinitely many sets. The following is a generalization of the associative laws.

Proposition 2.3.8. *Let $(A_i)_{i \in I}$ and $(B_i)_{i \in I}$ be indexed families of sets:*

1. $\bigcup_{i \in I} (A_i \cup B_i) = (\bigcup_{i \in I} A_i) \cup (\bigcup_{i \in I} B_i)$;
2. $\bigcap_{i \in I} (A_i \cap B_i) = (\bigcap_{i \in I} A_i) \cap (\bigcap_{i \in I} B_i)$.

Proof. (1) ($\subseteq$): Suppose that $x \in \bigcup_{i \in I} (A_i \cup B_i)$. Then, $(\exists i)\, x \in A_i \cup B_i$. Choose some k such that $x \in A_k \cup B_k$. Now, either $x \in A_k$ or $x \in B_k$. Without loss of generality suppose that $x \in A_k$. Then, $(\exists i)\, x \in A_i$, and therefore $x \in \bigcup_i A_i$. It follows that $x \in (\bigcup_{i \in I} A_i) \cup (\bigcup_{i \in I} B_i)$.

Note: When we say here "without loss of generality", we mean that a similar argument will work in the other case that $x \in B_k$.

($\supseteq$): Suppose now that $x \in (\bigcup_{i \in I} A_i) \cup (\bigcup_{i \in I} B_i)$. Then, either $x \in (\bigcup_{i \in I} A_i)$ or $x \in \cup(\bigcup_{i \in I} B_i)$. Without loss of generality suppose that $x \in \cup(\bigcup_{i \in I} B_i)$. Then, $(\exists i)\, x \in B_i$. Choose some k such that $x \in B_k$. Then, $x \in A_k \cup B_k$. Hence, $(\exists i)\, x \in A_i \cup B_i$. It follows that $x \in \bigcup_{i \in I} (A_i \cup B_i)$.

Part (2) is left to the exercises. $\qquad\qquad\square$

Here are some versions of the distributive law.

Proposition 2.3.9. *Let $(A_i)_{i \in I}$ and $(B_i)_{i \in I}$ be indexed families of sets:*

1. $A \cap \bigcup_{i \in I} B_i = \bigcup_{i \in I} (A \cap B_i);$
2. $A \cup \bigcap_{i \in I} B_i = \bigcap_{i \in I} (A \cup B_i).$

Proof. (1) ($\subseteq$): Let $x \in A \cap \bigcup_{i \in I} B_i$. Then, $x \in A$ and $x \in \bigcup_{i \in I} B_i$. Now, $(\exists i)\, x \in B_i$, so we can choose j such that $x \in B_j$. It follows that $x \in A \ \wedge \ x \in B_j$, so that $x \in A \cap B_j$. Thus, $(\exists i)\, x \in A \cap B_i$. Hence, $x \in \bigcup_{i \in I} A \cap B_i$.

($\supseteq$): Let $x \in \bigcup_{i \in I} A \cap B_i$. Then, $(\exists i)\, x \in \bigcup_{i \in I} A \cap B_i$, so we can choose j such that $x \in A \cap B_j$. Then, $x \in A \ \wedge \ x \in B_j$. Now, $(\exists i)\, x \in B_i$ so that $x \in \bigcup_{i \in I} B_i$. Thus, $x \in A \ \wedge \ x \in \bigcup_{i \in I} B_i$ so that $x \in A \cap \bigcup_{i \in I} B_i$.

Part (2) is left to the exercises. $\qquad\square$

Proposition 2.3.10. *Let $(A_i)_{i \in I}$ and $(B_i)_{i \in I}$ be indexed families of sets:*

1. $\bigcup_{i \in I} (A_i \cap B_i) \subseteq (\bigcup_{i \in I} A_i \cap \bigcup_{i \in I} B_i);$
2. $(\bigcap_{i \in I} A_i \cup \bigcap_{i \in I} B_i) \subseteq \bigcap_{i \in I} (A_i \cup B_i).$

Proof. Part (1) is left to the exercises. Here is a proof of Part (2). Let $x \in \bigcap_{i \in I} A_i \cup \bigcap_{i \in I} B_i$. Then either $x \in \bigcap_{i \in I} A_i$ or $x \in \bigcap_{i \in I} B_i$. Without loss of generality, suppose that $x \in \bigcap_{i \in I} A_i$. This means that $(\forall i \in I)\, x \in A_i$. Now, let $i \in I$ be arbitrary. Then, $x \in A_i$ so that $x \in A_i \ \vee \ x \in B_i$ and hence $x \in A_i \cup B_i$. Since i was arbitrary, it follows that $(\forall i \in I)\, x \in A_i \cup x \in B_i$. Thus, $x \in \bigcap_{i \in I} (A_i \cup B_i)$. $\quad\square$

Here is an example to show that equality does not always hold for the second inclusion. Let $I = \mathbb{Z}$, let A_i be the interval $(-\infty, i)$, and let $B_i = [i, \infty)$. Then $\bigcap_{i \in I} A_i = \emptyset = \bigcap_i B_i$, so that $\bigcap_{i \in I} A_i \cup \bigcap_{i \in I} B_i = \emptyset$. But $A_i \cup B_i = (-\infty, \infty)$ for every i, so that $\bigcap_{i \in I} (A_i \cup B_i) = (-\infty, \infty)$.

Finally, here is a version of DeMorgan's laws for indexed families.

Proposition 2.3.11. *Let $(A_i)_{i \in I}$ be an indexed family of sets:*

1. $(\bigcup_{i \in I} A_i)^{\complement} = \bigcap_{i \in I} A_i^{\complement};$
2. $(\bigcap_{i \in I} A_i)^{\complement} = \bigcup_{i \in I} A_i^{\complement}.$

Proof. (1) $x \in \left(\bigcup_{i \in I} A_i \right)^{\complement}$ if and only if $x \notin \bigcup_{i \in I} A_i$, which holds if and only if $\neg(\exists i)\, x \in A_i$. It follows from predicate logic that this is true if and only if $(\forall i)\, \neg x \in A_i$, which holds if and only if $(\forall i)\, x \in A_i^{\complement}$, which is true if and only if $x \in \bigcap_{i \in I} A_i^{\complement}$.

Part (2) is left to the exercises. $\qquad \square$

One can also define a doubly indexed family $\{B_{i,j} : i \in I, j \in J\}$. For example, let $A_{i,j}$ be the open interval $(i - j, i + j)$ of reals for $i \in \mathbb{Z}$ and $j \in \mathbb{N}$. Then we have

$$\bigcap_{i \in \mathbb{Z}} \bigcup_{j \in \mathbb{N}} A_{i,j} = \bigcap_{i \in \mathbb{Z}} \mathbb{R} = \mathbb{R},$$

whereas

$$\bigcup_{j \in \mathbb{N}} \bigcap_{i \in \mathbb{Z}} A_{i,j} = \bigcup_{j \in \mathbb{N}} \emptyset = \emptyset.$$

On the other hand, we do have the following.

Proposition 2.3.12. *For any doubly indexed family $\{A_{i,j} : i \in I, j \in J\}$, $\bigcup_{j \in J} \bigcap_{i \in I} A_{i,j} \subseteq \bigcap_{i \in I} \bigcup_{j \in J} A_{i,j}$.*

Proof. Suppose $x \in \bigcup_{j \in J} \bigcap_{i \in I} A_{i,j}$ and let $i \in I$ be arbitrary. Then $(\exists j \in J)\, x \in \bigcap_{i \in I} A_{i,j}$. Fix $k \in J$ such that $x \in \bigcap_{i \in I} A_{i,k}$. This means that $(\forall i \in I)\, x \in A_{i,k}$. Now, let $i \in I$ be arbitrary. Then we have immediately $x \in A_{i,k}$. Hence, $(\exists j)\, x \in A_{i,j}$ so that $x \in \bigcup_{j \in J} A_{i,j}$. Since i was arbitrary, it follows that $(\forall i \in I)\, x \in \bigcup_{j \in J} A_{i,j}$. This means that $x \in \bigcap_{i \in I} \bigcup_{j \in J} A_{i,j}$, as desired. $\qquad \square$

Exercises for Section 2.3

Exercise 2.3.1. Show that for any function $F : C \to D$ and any subsets A, B of D, $F^{-1}[A \setminus B] = F^{-1}[A] \setminus F^{-1}[B]$.

Exercise 2.3.2. Show that for any two functions $F : B \to C$ and $G : A \to B$, $F \circ G$ is a function.

Exercise 2.3.3. Show that for any functions F and G, $Dmn(F \circ G) = G^{-1}[Dmn(F)] \subseteq Dmn(G)$.

Exercise 2.3.4. Show that for any two functions F and G, $Rng(F \circ G) = F[Rng(G)] \subseteq Rng(F)$.

Exercise 2.3.5. Show that for any function $F : A \to B$, F is surjective if and only if, for all C and all $G : B \to C$ and $H : B \to C$, $G \circ F = H \circ F$ implies $G = H$.

Exercise 2.3.6. For a function $F : A \to B$, show the following:

1. F is injective if and only if there exists $G : B \to A$ such that $G \circ F = I_A$.
2. F is bijective if and only if there exists $G : B \to A$ such that $F \circ G = I_B$ and $G \circ F = I_A$.

Exercise 2.3.7. Let A, B, and C be sets:

(a) Show that $(A \cap B)^C = A^C \cap B^C$.
(b) Show that $A^C \cup B^C \subseteq (A \cup B)^C$.
(c) Show that equality does not always hold in (b).

Exercise 2.3.8. Let $A \subseteq B$:

(a) Show that $A^C \subseteq B^C$.
(b) Define a map from C^B onto C^A.

Exercise 2.3.9. Let $A_n = \{k/n : k \in \mathbb{Z}\} = \{0, 1/n, -1/n, 2/n, \dots\}$ for each $n \in \mathbb{N}^+$. Determine the resulting sets $\bigcup_{n \in I} A_n$ and $\bigcap_{n \in I} A_n$.

Exercise 2.3.10. Show that for any indexed families $\{A_i : i \in I\}$ and $\{B_i : i \in I\}$, $\bigcap_{i \in I}(A_i \cap B_i) = (\bigcap_{i \in I} A_i \cap \bigcap_{i \in I} B_i)$.

Exercise 2.3.11. Show that for any set A and any indexed family $\{B_i : i \in I\}$, $A \cup \bigcap_{i \in I} B_i = \bigcap_{i \in I}(A \cup B_i)$.

Exercise 2.3.12.

(a) Show that for any indexed families $\{A_i : i \in I\}$ and $\{B_i : i \in I\}$, $\bigcup_{i \in I}(A_i \cap B_i) \subseteq (\bigcup_{i \in I} A_i \cap \bigcup_{i \in I} B_i)$.
(b) Give an example to show that equality does not always hold in part (a).

Exercise 2.3.13. Let $\{A_i : i \in I\}$ and $\{B_j : j \in J\}$ be indexed families of sets and suppose that $A_i \subset B_j$ for all $i \in I$ and all $j \in J$. Show that $\bigcup_{i \in I} A_i \subset \bigcap_{j \in J} B_j$.

Exercise 2.3.14. Show that for any indexed family $\{A_i : i \in I\}$ of sets, $(\bigcap_{i \in I} A_i)^C = \bigcup_{i \in I} A_i^C$.

Exercise 2.3.15. Let $\{A_i : i \in I\}$ be an indexed family of sets:

(a) Show that $F[\bigcup_{i \in I} A_i] = \bigcup_{i \in I} F[A_i]$.
(b) Show that $F[\bigcap_{i \in I} A_i] \subset \bigcap_{i \in I} F[A_i]$.
(c) Show that equality does not always hold in (b).

Exercise 2.3.16. Let $\{B_i : i \in I\}$ be an indexed family of sets:

(a) Show that $F^{-1}[\bigcup_{i \in I} B_i] = \bigcup_{i \in I} F^{-1}[B_i]$.
(b) Show that $F^{-1}[\bigcap_{i \in I} B_i] = \bigcap_{i \in I} F^{-1}[B_i]$.

Exercise 2.3.17. Suppose $\mathcal{A} = \{A_i : i \in I\}$ is an indexed family, and $A_i \neq \emptyset$ for all $i \in I$. Show that the $Rng(p_i) = A_i$, where p_i is the i-th projection function on $\prod_{i \in I} A_i$.

Exercise 2.3.18. Let $I = \{0, 1\}$. Define a bijection between $\prod_{i \in I} A_i$ and $A_0 \times A_1$. More generally, define a bijection between $\prod_{i=1}^{n} A_i$ and $A_1 \times A_2 \times \cdots \times A_n$.

2.4 Equivalence Relations

Definition 2.4.1. A relation R on a set A is an *equivalence relation* if it is reflexive, symmetric, and transitive. For any $a \in A$, the *equivalence class* of a is $[a]_R := \{x \in A : aRx\}$, or sometimes written a/R.

A family $\{P_i : i \in I\}$ of subsets of A is a *partition* of A when the sets P_i are non-empty, pairwise disjoint, and $\bigcup_{i \in I} P_i = A$. The last two conditions may be rephrased to say that each element of A belongs to exactly one of the sets P_i.

Here are some well-known examples.

Example 2.4.2. For any positive integer m and any integers x, y, let $x \equiv y \, (mod \ m)$ if and only if m divides $x - y$. For example, if $m = 3$, then there are three equivalence classes, $[0] = \{0, 3, 6, \dots\}$, $[1] = \{1, 4, 7, \dots\}$, and $[2] = \{2, 5, 8, \dots\}$. The equivalence classes form the group $\mathbb{Z}(3)$ with addition take modulo 3.

Example 2.4.3. Let $F \equiv G$ (modulo finite) for functions $F, G : \mathbb{N} \to \mathbb{N}$ if and only if $\{x : F(x) \neq G(x)\}$ is finite. For subsets A, B of $\mathbb{N}$, let $A \equiv B$ (modulo finite) if and only if $\chi_A \equiv \chi_B$. Another way to phrase this is that $A \equiv B$ (modulo finite) if and only if the symmetric difference $(A \setminus B) \cup (B \setminus A)$ is finite.

The following proposition gives some key properties of equivalence classes.

Proposition 2.4.4. *Let R be an equivalence relation on A and let $[a]$ denote $[a]_R$. Then the following are equivalent:*

1. aRb;
2. $[a] = [b]$;
3. $a \in [b]$;
4. $[a] \cap [b] \neq \emptyset$.

Proof. $(1) \implies (2)$: Suppose aRb and let $c \in A$ be arbitrary. If $c \in [a]$, then cRa. By transitivity, this implies cRb and hence $c \in [b]$. Similarly, $c \in [b]$ implies $c \in [a]$. This demonstrates that $[a] = [b]$.

$(2) \implies (3)$: Suppose that $[a] = [b]$. Since $a \in [a]$, this implies that $a \in [b]$.

$(3) \implies (4)$: Suppose that $a \in [b]$. Since $a \in [a]$, this implies that $[a] \cap [b] \neq \emptyset$.

$(4) \implies (1)$: Suppose that $[a] \cap [b] \neq \emptyset$ and let $c \in [a] \cap [b]$. Then $c \in [a]$ so that aRc and $c \in [b]$ so that cRb. It follows from transitivity that aRb. $\square$

Proposition 2.4.5. *Let R be an equivalence relation on A, and $A \neq \emptyset$. Then the family of equivalence classes of A/R is partition of A.*

Proof. Let $[a]$ denote $[a]_R$. Certainly, each $[a]$ is a non-empty subset of A. $\bigcup_{a \in A}[a] = A$ since each $a \in [a]$. Given two classes $[a] \neq [b]$, it follows from Proposition 2.4.4 that $[a] \cap [b] = \emptyset$. $\square$

Proposition 2.4.6. *For any partition $P = \{A_i : i \in I\}$ of a set A, there is an equivalence relation R on A such that P is the set of equivalence classes of R.*

Proof. Define aRb if and only if a and b belong to the same set A_i of the partition. This is clearly reflexive and symmetric. Suppose now that aRb and bRc. Then for some $i, j \in I$, $a, b \in A_i$ and $b, c \in A_j$. Then $b \in A_i \cap A_j$ and therefore $A_i = A_j$ by the definition of partition. Thus, aRc. $\square$

Here is one general way to obtain an equivalence relation. The proof is left as an exercise.

Proposition 2.4.7. *Let $F : A \to B$ and define a relation R such that for all $x, y \in A$, $xRy \iff F(x) = F(y)$. Then R is an equivalence relation on A and there is a function $G : A/R \to B$ such that $F(a) = G([a]_R)$ for all $a \in A$.*

For example, let $F : \mathcal{P}(\mathbb{N}) \to \mathbb{N} \cup \{\omega\}$ be defined by $F(A) = card(A)$, where $card(A)$ is the cardinality of the set A. Here we write ω for the cardinality of $\mathbb{N}$. Then the resulting equivalence relation will have $A \equiv B$ if and only if A and B have the same cardinality. For instance, the equivalence class of $A = \{3\}$ is $[A] = \{\{0\}, \{1\}, \dots\}$, that is, the family of sets having exactly one element. The function G thus satisfies $G([A]) = card(A)$.

Exercises for Section 2.4

Exercise 2.4.1. Suppose we extend the equivalence relation of equality modulo k to the real numbers, meaning that $x \equiv y \, (mod \ k)$ if $x - y$ is an integer multiple of k. Prove that this is still an equivalence relation. Find the members of the equivalence class of the real number π, when $k = 2$. What can you say about the group $\mathbb{R} \, (mod \ k)$?

Exercise 2.4.2. What is the equivalence class of $\emptyset$ under equivalence modulo finite?

Exercise 2.4.3. Let $F : A \to B$ and define a relation R so that for all $x, y \in A$, $xRy \iff F(x) = F(y)$. Show that R is an equivalence relation on A and that there is a function $G : A/R \to B$ such that $F(a) = G(a/R)$ for all a.

Exercise 2.4.4.

(a) Let R and S be equivalence relations on a non-empty set A, and suppose $R \subseteq S$. Show that S induces an equivalence relation T on A/R given by $[a]_R T [b]_R \iff aSb$.
(b) Let R be an equivalence relation on A and let T be an equivalence relation on A/R. Show that T induces an equivalence relation S on A such that $R \subseteq S$ given by $aSb \iff [a]_R T [b]_R$.

Exercise 2.4.5. Let R be an equivalence relation on A and let $B \subseteq A$:

(a) Show that $R \cap (B \times B)$ is an equivalence relation on B.
(b) Let $F : B \to A$. Show that F induces an equivalence relation S on B, given by $aSb \iff F(a)RF(b)$.

Exercise 2.4.6. Show that if $\{R_i : i \in I\}$ is a family of equivalence relations on a set A and then $\bigcap_{i \in I} R_i$ is also an equivalence relation on A.

Exercise 2.4.7. Let R and S be two equivalence relations on a set A. Show that $R \subseteq S$ if and only if every equivalence class K of R is included in some equivalence class M of S.

2.5 Orderings

In this section, we introduce the various types of partial and total orderings.

Definition 2.5.1. A relation R on A is a *preorder* or *quasiorder* if it is reflexive and transitive. A preorder is a *partial ordering* if it is antisymmetric. If A is a set equipped with a partial ordering, we refer to A as a *partially ordered set*, or *p.o. set* for short. A partial ordering R is a *linear ordering* if it is *total*, that is, for any $a, b \in A$, either aRb or bRa. An ordering R is a *well-ordering* if it is linear and *well-founded*, that is, any subset B of A has a least element. A totally ordered subset of a partially ordered set is called a *chain*.

Note that a preorder which is symmetric is just an equivalence relation. The relation aRb if and only if $card(a) \leq card(b)$ for sets a and b is an example of a preorder.

Given a set A partially ordered by $\leq$ and a subset B of A, we write $a < b$ for $a \leq b$ & $a \neq b$. The relation $<$ is irreflexive antisymmetric and transitive. More generally, R is a *strict partial ordering* if it is irreflexive, antisymmetric, and transitive. Another way to say that a partial order is total, or linear, is by the following.

Definition 2.5.2. The partial order $\leq$ on a set A has the *Trichotomy Property* if, for any $x, y \in A$, exactly one of the following holds:

1. $x < y$;
2. $y < x$;
3. $x = y$.

The standard ordering $\leq$ of the real numbers is a linear ordering, and the corresponding strict order is given by $<$.

The notions of minimal and maximal elements in a partially ordered set, as well as lower and upper bounds for ordered sets, are very important.

Definition 2.5.3. Let $\leq$ be a partial ordering on a set A and let $B \subseteq A$:

1. a is a *minimal* element of B if $a \in B$ and there is no $b \in B$ with $b < a$.
2. a is a *maximal* element of B if $a \in B$ and there is no $b \in B$ with $b > a$.
3. a is the *minimum* element of B if $a \in B$ and for every $b \in B$, $a \leq b$.
4. a is the *maximum* element of B if $a \in B$ and for every $b \in B$, $a \leq b$.
5. a is a *lower bound* for B if $a \leq b$ for every $b \in B$. Here a does not have to belong to B.
6. a is an *upper bound* for B if $b \leq a$ for every $b \in B$. Again a does not have to belong to B.
7. a is the *greatest lower bound* or *infimum* of B if a is a lower bound and $c \leq a$ for every lower bound c of B.
8. a is the *least upper bound* or *supremum* of B if a is an upper bound and $a \leq c$ for every upper bound c of B.

We consider some examples of partial orderings and related concepts involving divisibility for natural numbers, inequality for real numbers, and inclusion for sets.

Example 2.5.4.

1. Consider the divisibility relation on the positive integers. This relation is reflexive, since $x = x \cdot 1$, so $x \mid x$ for any x. To see that it is transitive, suppose $a \mid b$ and $b \mid c$. Then $b = ax$ and $c = by$ for some positive integers x, y. It follows that xy is also a positive

integer and $c = axy$. To see that it is antisymmetric, suppose that $a|b$ and $b|a$. Then $b = ax$ and $a = by$ for some positive integers x, y. Then xy is a positive integer and $a = axy$. It follows that $xy = 1$ and therefore $x = y = 1$ and $a = b$. Note that we have $a \mid -a$ and $-a \mid a$, so the divisibility relation is not antisymmetric on the integers.

2. For two positive integers a, b, the maximum and the minimum under divisibility exist if one of the two divides the other and then these are just the usual $\max\{a, b\}$ and $\min\{a, b\}$ under the standard ordering $\leq$. The least upper bound of $\{a, b\}$ is the least common multiple and the greatest lower bound of $\{a, b\}$ is the greatest common denominator.

3. Let $\leq$ be the standard ordering on the real numbers. A finite closed interval $[a, b]$ has minimal element a and maximal element b. A finite open interval (a, b) has infimum a and supremum b but has no minimal or maximal elements. The half-open interval $[0, \infty)$ has no supremum. The set $\{1/n : n \in \mathbb{N}^+\}$ has infimum 0 but has no minimum.

4. For the inclusion relation on $\mathcal{P}(\mathbb{N})$, let B be the family of non-empty sets. Then for any $n \in \mathbb{N}$, the singleton $\{n\}$ is a minimal element of B, but B has no minimum element.

Exercises for Section 2.5

Exercise 2.5.1. Show that if $\leq$ is a partial ordering with strict order $<$, then it is linear if and only if it satisfies the Trichotomy Property.

Exercise 2.5.2. Show that if R is a preordering, then R^{-1} is a preordering.

Exercise 2.5.3. Show that if R is a preordering, then $R \cap R^{-1}$ is an equivalence relation.

Exercise 2.5.4.

(a) Show that if a is the minimum element in a subset B of a p.o. set A, then a is the unique minimal element of B.

(b) Give an example of a p.o. set with a unique minimal element but no minimum element.

Exercise 2.5.5.

(a) Show that a maximum element of a subset B of a p.o. set is always the supremum of B.

(b) Show that the supremum a of B is the maximum of B if and only if $a \in B$.

Exercise 2.5.6. Show that any well-founded partial ordering is a well-ordering. That is, show that such an ordering must be totally ordered.

Exercise 2.5.7. Prove that for any p.o. set A, there is an injection $F : A \to \mathcal{P}(A)$ such that $a \leq b \iff F(A) \subseteq F(B)$.

Exercise 2.5.8. Let $(B, \leq)$ be a set with a partial ordering. Let $F : A \to B$ and define a relation R on $Dmn(F)$ by $xRy \iff F(x) \leq F(y)$. Show that R is a preorder.

Exercise 2.5.9. Let $F : (A, \leq_A) \to (B, \leq_B)$ map one linearly ordered set onto another so that $a_1 < a_2$ implies $F(a_1) < F(a_2)$. Show that F is an order isomorphism, that is, F is one-to-one and $a_1 \leq a_2 \iff F(a_1) \leq F(a_2)$.

Exercise 2.5.10. Given a preorder R on a set A, prove that there is an equivalence relation S on A and a partial ordering $\leq$ on A/S such that $[a]_S \leq [b]_S \iff aRb$.

2.6 Trees

Trees play a very important role in many areas of mathematics, set theory, and logic in particular. There are many ways to present the notion of a tree. In graph theory, a tree is a connected, acyclic directed graph; that is, there is no *cycle* of directed edges $(v_0, v_1), (v_1, v_2), \ldots, (v_{n-1}, v_n), (v_n, v_0)$.

We are interested in *rooted* trees, where there is a special node called the root and a directed path from the root to every node, where a *directed path* from node v_0 to node v_n is a sequence $(v_0, v_1), (v_1, v_2), \ldots, (v_{n-1}, v_n)$ of directed edges. (Note that the empty sequence may be viewed as a path from v to itself.)

When there is a directed edge from u to v, we say that v is a *successor* of u and v is the (unique) predecessor of v. There is a natural partial ordering on a tree, defined so that $u \leq v$ if and only if there is a path from u to v.

Example 2.6.1. For example, let $T = \mathbb{N}$ and let mEn if and only if $n = m + 1$. Then 0 is the root and each number m has unique successor $m + 1$. The partial ordering given by this tree is simply the standard $\leq$ ordering on $\mathbb{N}$.

We want to consider trees with *strings* as vertices. Let Σ be a set of symbols (an *alphabet*), usually an initial segment of $\mathbb{N}$. Then for a natural number n, Σ^n denotes the set of strings $\sigma = (\sigma(0), \sigma(1), \dots, \sigma(n-1))$ of n letters from Σ; the length n of σ is denoted by $|\sigma|$. The empty string has length 0 and will be denoted by ϵ. Σ^* (or sometimes $\Sigma^{<\omega}$) denotes the set $\bigcup_{n \in \mathbb{N}} \Sigma^n$ and Σ^ω, or $\Sigma^{\mathbb{N}}$, denotes the set of infinite sequences.

A constant string σ of length n consisting of the symbol k is denoted k^n. For $m < |\sigma|$, $\sigma \restriction m$ is the string $(\sigma(0), \dots, \sigma(m-1))$; σ is an *initial segment* of τ (written $\sigma \sqsubseteq \tau$) if $\sigma = \tau \restriction m$ for some m. Initial segments are also referred to as *prefixes*. Similarly, τ is said to be a *suffix* of σ if $|\tau| \leq |\sigma|$ and, for all $i < |\tau|$, $\sigma(|\sigma| - |\tau| + i) = \tau(i)$. The *concatenation* $\sigma^\frown \tau$ (or sometimes $\sigma * \tau$ or just $\sigma\tau$) is defined by $\sigma^\frown \tau = (\sigma(0), \sigma(1), \dots, \sigma(m-1), \tau(0), \tau(1), \dots, \tau(n-1))$, where $|\sigma| = m$ and $|\tau| = n$; in particular, we write $\sigma^\frown a$ for $\sigma^\frown (a)$ and $a^\frown \sigma$ for $(a)^\frown \sigma$. Thus, we may also say that σ is a prefix of τ if and only if $\tau = \sigma^\frown \rho$ for some ρ and that τ is a suffix of σ if and only if $\sigma = \rho^\frown \tau$ for some ρ.

For any $x \in \Sigma^{\mathbb{N}}$ and any finite n, the *initial segment* $x \restriction n$ of x is $(x(0), \dots, x(n-1))$. We write $\sigma \sqsubseteq x$ if $\sigma = x \restriction n$ for some n. For any $\sigma \in \Sigma^n$ and any $x \in \Sigma^{\mathbb{N}}$, we have $\sigma^\frown x = (\sigma(0), \dots, \sigma(n-1), x(0), x(1), \dots)$.

Proposition 2.6.2. *The relation $\sqsubseteq$ is a partial ordering.*

Proof. The reflexive property is immediate from the definition.

For antisymmetry, suppose that $\sigma \sqsubseteq \tau$ and $\tau \sqsubseteq \sigma$. Then $\tau = \sigma^\frown \mu$ and $\sigma = \tau^\frown \nu$ for some strings μ and ν. But this means $\tau = \sigma^\frown \mu^\frown \nu$ so that $\mu = \epsilon = \nu$ and therefore $\sigma = \tau$.

For transitivity, suppose $\sigma \sqsubseteq \tau$ and $\tau \sqsubseteq \rho$. Then $\tau = \sigma^\frown \mu$ and $\rho = \tau^\frown \nu$ for some strings μ and ν.

Therefore, $\rho = \sigma^\frown \mu^\frown \nu$ so that $\sigma \sqsubseteq \rho$. $\qquad\square$

Definition 2.6.3. The *lexicographic* or dictionary ordering on Σ^* is defined so that $\sigma \preceq \tau$ if $\sigma \sqsubseteq \tau$ or if $\sigma(n) < \tau(n)$, where n is the least such that $\sigma(n) \neq \tau(n)$.

When $\sigma \preceq \tau$ but $\sigma \neq \tau$, we write $\sigma \prec \tau$.

Proposition 2.6.4. *The relation $\preceq$ is a linear ordering.*

Proof. *Reflexive*: $\sigma \subseteq \sigma$ so that $\sigma \preceq \sigma$.

Antisymmetric: Suppose that $\sigma \preceq \tau$ and $\tau \preceq \sigma$. Observe that if, for some n, $\sigma \restriction n = \tau \restriction n$ and $\sigma(n) < \tau(n)$, then it not true that $\tau \preceq \sigma$. It follows that $\sigma \sqsubseteq \tau$ and, similarly, $\tau \sqsubseteq \sigma$ so that $\sigma = \tau$.

Transitive: Suppose that $\sigma \preceq \tau$ and $\tau \preceq \rho$. There are two cases:

Case 1: $\sigma \subseteq \tau$. Now, there are two subcases.
In the first subcase, $\tau \sqsubset \rho$, then $\sigma \sqsubset \rho$ by Proposition 2.6.2 and therefore $\sigma \preceq \rho$.

In the second subcase, $\tau \restriction n = \rho \restriction n$ and $\tau(n) < \rho(n)$ for some n. There are still two further possibilities:

(i) If $|\sigma| \leq n$, then $\sigma \sqsubseteq \rho$ and therefore $\sigma \preceq \rho$.

(ii) If $|\sigma| > n$, then $\sigma \restriction n = \tau \restriction n = \rho \restriction n$ and $\sigma(n) = \tau(n) < \rho(n)$ so that again $\sigma \preceq \rho$.

Case 2: There is some n such that $\sigma \restriction n = \tau \restriction n$ and $\sigma(n) < \tau(n)$. This is left as an exercise.

Total: Given two strings σ and τ, suppose without loss of generality that $|\sigma| \leq |\tau|$. There are two cases:

Case 1: $\sigma = \tau \restriction n$ so that $\sigma \sqsubset \tau$ and therefore $\sigma \preceq \tau$.

Case 2: For some $i < n$, $\sigma(i) \neq \tau(i)$. Let m be the least such i and assume without loss of generality that $\sigma(m) < \tau(m)$. Then $\sigma \preceq \tau$. $\square$

Finite strings in $\{0,1\}^*$ can be viewed as representing *dyadic* rational numbers between 0 and 1.

Definition 2.6.5. For $\sigma \in \{0,1\}^*$, let $q_\sigma = \sum_{i<|\sigma|} 2^{-i-1}\sigma(i)$.

For example, 1101 represents $q_{1101} = \frac{1}{2} + \frac{1}{4} + \frac{1}{16} = \frac{13}{16}$. Note here that the number 13 in base two is just 1101, that is, $13 = 8 + 4 + 1$. Note that 110100 also represents $\frac{13}{16}$. For a dyadic rational $q \in (0,1)$, the standard representation σ_q is the one that ends with a 1, and the standard representation of 0 is 0.

A similar definition can be given for numbers in base k when the alphabet $\Sigma = \{0, 1, \ldots, k-1\}$.

Proposition 2.6.6.

1. *For any σ and τ in $\{0,1\}^*$, if $\sigma \preceq \tau$, then $q_\sigma \leq q_\tau$.*
2. *For any dyadic rationals p and q in $[0, 1)$, $\sigma_p \preceq \sigma_q$ if and only if $p \leq q$ in the usual ordering of the rationals.*

Proof. The proof of part (1) is left as an exercise.

(2) Let p and q be rationals in $[0, 1]$ and let $\sigma = \sigma_p$ and $\tau = \sigma_q$ so that $p = q_\sigma$ and $q = q_\tau$.

Suppose first that $\sigma_p \preceq \sigma_q$. Then by part (1), we obtain $p = q_\sigma \leq q_\tau = q$.

Suppose next that $p \leq q$ and, by way of contradiction, suppose that it is not true that $\sigma \preceq \tau$. It follows from Proposition 2.6.4 that $\tau \preceq \sigma$. Then by the first case above, $q \leq p$. Thus, $p = q$ and therefore $\sigma_p = \sigma_q$. $\qquad\square$

A *tree* T over Σ is a set of finite strings from Σ^* which is closed under initial segments. Then $\tau \in T$ is an *immediate successor* of a string $\sigma \in T$ if $\tau = \sigma^\frown a$ for some $a \in \Sigma$.

We will generally assume that $T \subseteq \mathbb{N}^*$. That is, the nodes of T are finite sequences of natural numbers. Such a tree defines a subset $[T]$ of the so-called Baire space $\mathbb{N}^\mathbb{N}$, where $[T]$ is the set of infinite paths through the tree T. That is,

$$x \in [T] \iff (\forall n)\, x \upharpoonright n \in T.$$

Example 2.6.7. The full binary tree is $\{0,1\}^*$. Here every node σ has exactly two successors: $\sigma 0$ and $\sigma 1$. The set of infinite paths through the full tree is the so-called Cantor space $\{0,1\}^\mathbb{N}$.

A tree T is said to be a *shift* if it is also closed under suffixes.

Example 2.6.8. Define $T \subseteq \{0,1\}^*$ so that $\sigma \in T$ if and only if σ does not have 3 consecutive 0's, that is, if σ has no consecutive substring of the form (000). Clearly, if σ does not have 3 consecutive 0's, then no initial segment of σ can have 3 consecutive 0's either.

Furthermore, if σ has no consecutive substring (000), then no suffix of σ can have a consecutive substring (000). Thus, T is a shift.

We say that a tree T is *finite-branching* if for every $\sigma \in T$, there are only finitely many immediate successors of σ in T. Certainly, any tree T over a finite alphabet is finite-branching.

Example 2.6.9. Define the tree $T \subseteq \omega^*$ so that for any string $\sigma \in T$ and any $i < |\sigma|$, $\sigma(i) \leq i$. Then for any $\sigma \in T$ of length n, σ has at most $n + 1$ successors.

Exercises for Section 2.6

Exercise 2.6.1. For any σ and τ in $\{0, 1\}^*$, if $\sigma \preceq \tau$, then $q_\sigma \leq q_\tau$.

Exercise 2.6.2. Show that the lexicographic ordering on $\{0, 1\}^*$ is not well-founded.

Hint: The ordering on the dyadic rational numbers is not well-founded.

Exercise 2.6.3. Complete the proof that the lexicographic ordering on $\{0, 1\}^*$ is a linear ordering.

Exercise 2.6.4. Show that there is no cycle in the full tree ω^*.

Exercise 2.6.5. For any abstract tree $T = (V, E)$, and any node u of T, let $T(u)$ be the set of nodes v such that there is a path from u to v. Show that $T(u)$ is also a tree, that is, $T(u)$ does not contain any cycle.

Chapter 3

Zermelo–Fraenkel Set Theory

3.1 Historical Context

In the 19th century, mathematicians produced a great number of sophisticated theorems and proofs. With the increasing sophistication of their techniques, an important question appeared now and again: Which theorems require a proof and which facts are self-evident to a degree that no sensible mathematical proof of them is possible? What are the proper boundaries of mathematical discourse? The contents of these questions is best illustrated by several contemporary examples.

Example 3.1.1. The parallel postulate of Euclidean geometry was a subject of study for centuries. The study of geometries that fail to satisfy this postulate was considered a non-mathematical folly prior to the early 19th century, and Gauss withheld his findings in this direction for fear of public reaction. The hyperbolic geometry was discovered only in 1830 by Lobachevsky and Bolyai. Non-Euclidean geometries proved to be an indispensable tool in the general theory of relativity.

Example 3.1.2. The Jordan curve theorem asserts that every non-self-intersecting closed curve divides the Euclidean plane into two regions, one bounded and the other unbounded, and any path from the bounded to the unbounded region must intersect the curve. The proof was first presented in 1887. The statement sounds self-evident, but the initial proofs were found to be confusing and unsatisfactory.

The consensus formed that even statements of this kind must be proved from some more elementary properties of the real line.

Example 3.1.3. Georg Cantor produced an exceptionally simple proof of the existence of non-algebraic real numbers, i.e. real numbers which are not roots of any polynomial with integer coefficients (1874). Proving that specific real numbers such as π or e are not algebraic is quite difficult, and the techniques for such proofs were under development at that time. On the other hand, Cantor only compared the cardinalities of the sets of algebraic numbers and real numbers, found that the first has smaller cardinality, and concluded that there must be real numbers that are not algebraic without ever producing a single definite example. Cantor's methodology — comparing cardinalities of different infinite sets — struck many people as non-mathematical.

As a result, the mathematical community in the late 19th century experienced an almost universally acknowledged need for an axiomatic development of mathematics modeled after the classical axiomatic treatment of geometry by Euclid. It was understood that the primitive concept must be that of a set (as opposed to a real number, for example) since the treatment of real numbers can be fairly easily reinterpreted as speaking about sets of a certain specific kind. The need for a careful choice of axioms was accentuated by several paradoxes, of which the simplest and most famous is *Russell's paradox*: Consider the "set"x of all sets z which are not elements of themselves. Consider the question whether $x \in x$ or not. If $x \in x$, then x does not satisfy the formula used to define x, and so, $x \notin x$. On the other hand, if $x \notin x$, then x does satisfy the formula used to define x, and so, $x \in x$. In both cases, a contradiction appears. Thus, the axiomatization must be formulated in a way that avoids this paradox.

Several attempts at a suitable axiomatization appeared before Zermelo produced his collection of axioms in 1908, now known as the Zermelo set theory with choice (ZC). After a protracted discussion and two late additions, the axiomatization of set theory stabilized in the 1920s in the form now known as Zermelo–Fraenkel set theory, ZF, and ZF with the Axiom of Choice (ZFC). We note that the first modern textbook on axiomatic set theory was written by Abraham Fraenkel [4]. This process finally placed mathematics on a strictly formal foundation. A mathematical statement is one that can be faithfully represented as a formula in the language of set theory. A correct mathematical argument is one that can be rewritten as a formal proof from the axioms of ZFC. Here (roughly), a formal proof of a formula ϕ from the axioms is a finite sequence of formulas ending with ϕ such that each formula on the sequence is either one of the axioms or follows from the previous formulas on the sequence using a fixed collection of formal derivation rules.

The existence of such a formal foundation does not mean that mathematicians actually bother to strictly conform to it. Russell and Whitehead's *Principia Mathematica* [9] was a thorough attempt to rewrite many mathematical arguments in a formal way, using a theory different from ZFC. It showed among other things that a purely formal treatment is excessively tiresome and adds very little insight. Long, strictly formal proofs of mathematical theorems of any importance have been produced only after the advent of computers. Mathematicians still far prefer to verify their argument by social means, such as by presentations at seminars or conferences or in publications. The existence of a strictly formal proof is considered as an afterthought and a mechanical consequence of the existence of a proof that conforms to the present socially defined standards of rigor. In this treatment, we will also produce non-formal rigorous proofs in ZFC with the hope that the reader can accept them and learn to emulate them.

3.2 The Language of the Theory

Zermelo–Fraenkel set theory with the Axiom of Choice (ZFC) belongs to the class of theories known as first-order theories. General first-order theories will be developed in the companion volume [3]. Here, we only look at the special case of ZFC. The language of set theory consists of the following symbols:

- an infinite supply of variables;
- a complete supply of logical connectives: implication $\rightarrow$, conjunction $\wedge$, disjunction $\vee$, equivalence $\leftrightarrow$, and negation $\neg$;
- quantifiers: the universal quantifier $\forall$ (read "for all") and the existential quantifier $\exists$ (read "there exists");
- equality $=$;

- the binary relational symbol $\in$ (the intended interpretation of which is membership; read "belongs to", "is an element of").

The symbols of the language can be used in prescribed ways to form expressions which we refer to as formulas. In the case of ZFC, if x and y are variables, then $x = y$ and $x \in y$ are formulas; if ϕ, ψ are formulas, then so are $\phi \wedge \psi$, $\neg\phi$, etc.; and if ϕ is a formula and x is a variable, then $\forall x\ \phi$ and $\exists x\ \phi$ are formulas; in these formulas, ϕ is called the scope of the quantifier $\forall x$ or $\exists x$ (we will sometimes write quantifiers as $(\forall x)$ or $(\exists x)$ for the sake of readability). Formulas are customarily denoted by Greek letters, such as ϕ, ψ, θ. A variable x is *free* in a formula ϕ if it appears in ϕ outside of scope of any quantifier. Often, the free variables of a formula are listed in parentheses: $\phi(x)$, $\psi(x, y)$. A formula with no free variables is called a *sentence*.

Even quite short formulas in this rudimentary language tend to become entirely unreadable. To help understanding, mathematicians use a great number of shorthands, which are definitions of certain objects or relations among them. Among the most common shorthands in ZFC are the following:

- $\forall x \in y\ \phi$ is a shorthand for $(\forall x)[\, x \in y \to \phi\,]$, and $\exists x \in y\ \phi$ is a shorthand for $(\exists x)[\, x \in y \wedge \phi\,]$;
- $\exists! x\ \phi$ is short for "there exists exactly one", in other words for

$$(\exists x)[\, \phi(x) \ \wedge \ (\forall y)[\, \phi(y) \to y = x\,]\,];$$

- $x \subseteq y$ (subset) is short for $\forall z[\, z \in x \to z \in y\,]$;
- $\emptyset$ and also 0 are shorthands for the empty set (the unique set with no elements);
- $x \cup y$ and $x \cap y$ denote the union and intersection of sets x, y;
- $\mathcal{P}(x)$ denotes the power set of x, the set of all subsets of x.

After the development of functions, arithmetical operations, real numbers, etc., more shorthands appear, including the familiar $\mathbb{R}$, $+$, $\sin x$, $\int f(x)dx$ and so on. Any formal proof in ZFC using these shorthands can be mechanically rewritten into a form which does not use them. Since the shorthands really do make proofs shorter and easier to understand, we will use them whenever convenient.

3.3 The Basic Axioms

At the basis of any first-order theory, there is a body of axioms known as the *logical axioms*. They record the behavior of the underlying logic and have nothing to do with the theory *per se*. The choice of logical axioms depends on the precise definition of the formal proof system one wants to use. They are typically statements like the following: $(\forall x)[\, x = x\,]$, $(\forall x)(\forall y)(\forall z)[[\, x = y \wedge y = z\,] \to x = z\,]$, or $\phi \to (\psi \to \phi)$ for any formulas ϕ, ψ. The possible choices for the system of the logical axioms are discussed in the companion book, *Foundations of Mathematics* [3]; we will not explain them here.

Any set may be determined by its elements. This is codified in the first axiom.

Definition 3.3.1. The *Extensionality Axiom* states that

$$(\forall x)(\forall y)[(\forall z)[z \in x \leftrightarrow z \in y\,] \to x = y\,].$$

In other words, two sets with the same elements are equal. We used this idea in the first chapter when we said that two sets A and B are equal if each is a subset of the other.

Several of the axioms of ZFC are needed to provide the existence of basic sets and Boolean operations on sets.

Definition 3.3.2. The *Empty Set Axiom* asserts the existence of the empty set, that is, $(\exists x)(\forall y)[\, y \notin x\,]$.

The empty set is unique by the Axiom of Extensionality and we will denote it by $\emptyset$ or just 0.

Definition 3.3.3. The *Pairing Axiom* says that for any two sets x and y, there is a set $\{x, y\}$, that is,

$$(\forall x)(\forall y)(\exists z)(\forall u)[\, u \in z \leftrightarrow [\, u = x \vee u = y\,]].$$

Writing the set $\{x, y\}$ is our first use of the *set builder notation*. Note that if $x = y$, we get a singleton set $\{x\}$. Here, this notation means that $x \in \{a\} \iff x = a$, and in general,

$$x \in \{a_1, \ldots, a_n\} \iff (x = a_1 \vee \cdots \vee x = a_n).$$

The pair $\{x, y\}$ is unordered: Looking at it, we cannot tell whether x comes first and y second or vice versa. We will also use the ordered pair (x, y), which (as a definition of Kuratowski) is the set $\{\{x\}, \{x, y\}\}$. This can be formed using the pairing axiom several times. Ordered triples would be defined as $(x, y, z) = ((x, y), z)$, similar to ordered n-tuples for every natural number n.

Proposition 3.3.4. *For any sets a, b, c, d, $\{a, b\} = \{c, d\} \iff (a = c \wedge b = d) \vee (a = d \wedge b = c)$.*

Proof. $(\Longleftarrow)$ Suppose that $a = c$ and $b = d$. Then

$$x \in \{a, b\} \iff (x = a \vee x = b) \iff (x = c \vee x = d) \iff x \in \{c, d\}.$$

The case when $a = d$ and $b = c$ is similar.

$(\Longrightarrow)$ Suppose that $\{a, b\} = \{c, d\}$. Since $a \in \{a, b\}$, it follows that $a \in \{c, d\}$. Thus, either $a = c$ or $a = d$. Without loss of generality, let $a = c$. By a similar argument, either $b = c$ or $b = d$. In the first case, $a = c = b$, and therefore, also $a = b = d$, since $d \in \{a, b\}$. In either case, we have $b = d$, so that $a = c$ and $b = d$. $\qquad\square$

Definition 3.3.5. The *Union Axiom* asserts the existence of the union $\bigcup x$ of any set of sets, that is,

$$(\forall x)(\exists y)(\forall z)[z \in y \leftrightarrow (\exists u)[u \in x \wedge z \in u]].$$

Note that the union of x is uniquely given by this description due to the Axiom of Extensionality. This is a generalization of the union of an indexed family discussed in Chapter 2. If $A = \{A_i : i \in I\}$ is an indexed family of sets, then $\bigcup A = \bigcup_{i \in I} A_i$. Given an arbitrary set A, let $I = A$ and let $A_i = \{i\}$. Then $\bigcup A = \bigcup_{i \in I} A_i$.

The union and pairing axioms make it possible to formulate several operations on sets. If x, y are sets, then there is a unique set (denoted by $x \cup y$) containing exactly elements of x and elements of y: $x \cup y = \bigcup\{x, y\}$. Given a finite list $x_0, x_1, x_2, \ldots, x_n$ of sets, we can form the set $\{x_0, x_1, x_2, \ldots x_n\}$, which is the unique set containing exactly the sets on our list.

We can now introduce the von Neumann natural numbers. These begin with $0 = \emptyset$ and are then recursively defined by $n + 1 = n \cup \{n\}$. Thus, $1 = 0 \cup \{0\} = \{0\}$, $2 = 1 \cup \{1\} = \{0\} \cup \{1\} = \{0, 1\}$, $3 = \{0, 1, 2\}$, and so on.

We note here that beginning with just the empty set and closing under the operations of pairing and union, we produce the family HF of *hereditarily finite* sets. This includes sets, such as $\{1, \{2, 4\}, \{\{7\}\}\}$.

Definition 3.3.6. The *Power Set Axiom* asserts that

$$(\forall x)(\exists y)(\forall z)[\, z \in y \leftrightarrow z \subseteq y\,].$$

In other words, for every set x, there is a set consisting exactly of all subsets of x. This set is called the *power set* of x and commonly denoted by $\mathcal{P}(x)$. Note that the power set of x is uniquely given by this description by the Axiom of Extensionality.

For example, since $2 = \{0, 1\}$, $\mathcal{P}(2) = \{0, \{0\}, \{1\}, \{0, 1\}\} = \{0, 1, \{1\}, 2\}$. It can be seen that the power set of any hereditarily finite set is also hereditarily finite.

Exercises for Section 3.3

Exercise 3.3.1. Does $0 = \{0\}$ hold? Why or why not?

Exercise 3.3.2. Evaluate $\bigcup \{0, \{0\}, \{0, \{0\}\}\}$ in the set builder notation.

Exercise 3.3.3. Show that (a, b) is a set for any sets a and b.

Exercise 3.3.4. Show that $(a, b) = (c, d)$ if and only if $a = c$ and $b = d$ for any sets a and b.

Exercise 3.3.5. Show that, for any sets x and y, if $x \in A \iff y \in A$ for all sets A, then $x = y$.

Exercise 3.3.6. Prove by induction on n that for any natural number n and any n sets $a_1, \ldots, a_n$, there exists a set containing exactly those n elements. Use only the Pairing and Union axioms.

Exercise 3.3.7. Prove by induction on n that $n = \{0, 1, \ldots, n-1\}$ for all natural numbers n.

Exercise 3.3.8. Write a formula $\phi(x, y, z)$ of set theory which says the following: "z is an ordered pair whose first element is x and the second element is y".

Exercise 3.3.9. Evaluate $\mathcal{P}(3)$ in the set builder notation.

Exercise 3.3.10. Show that for any sets A and B,

(a) if $A \subseteq B$, then $\mathcal{P}(A) \subseteq \mathcal{P}(B)$;
(b) $\mathcal{P}(A \cap B) = \mathcal{P}(A) \cap \mathcal{P}(B)$.

Exercise 3.3.11.

(a) Show that for any sets A and B,
  $\mathcal{P}(A) \cup \mathcal{P}(B) \subseteq \mathcal{P}(A \cup B)$.
(b) Give an example to show that equality does not always hold
  in (a).

Exercise 3.3.12. Show that for any sets A and B,

(a) if $A \subseteq B$, then $\bigcup A \subseteq \bigcup B$;
(b) $\bigcup(A \cup B) = \bigcup A \cup \bigcup B$.

Exercise 3.3.13.

(a) Show that for any sets A and B,
  $\bigcup(A \cap B) \subseteq \bigcup A \cap \bigcup B$.
(b) Give an example to show that equality does not always hold
  in (a).

Exercise 3.3.14. Show that $\bigcup \mathcal{P}(A) = A$ for any set A.

Exercise 3.3.15. Show that $A \subseteq \mathcal{P}(\bigcup A)$ for any set A. Find an example to show that equality does not always hold.

3.4 Axiom of Infinity

To move beyond the hereditarily finite sets, we need to assert the existence of an infinite set. The set we have in mind is $\mathbb{N} = \{0, 1, 2, \dots\}$, typically denoted in set theory as ω. Recall that in set theory, we have defined the natural numbers by $0 = \emptyset$ and $n + 1 = n \cup \{n\}$ for each n.

Definition 3.4.1. The *Axiom of Infinity* is the statement

$$(\exists x)[0 \in x \wedge (\forall y \in x)[y \cup \{y\} \in x]].$$

A brief discussion reveals that the set x in question must be in some naive sense infinite: Its elements are 0, $\{0\}$, $\{0, \{0\}\}$ and so on.

The intended infinite set is simply the set $\omega = \{0, 1, 2, \dots\}$ of natural numbers, also denoted by $\mathbb{N}$. This will be discussed in the following chapter. One must keep in mind that the distinction between finite and infinite sets must be defined formally. This is done in Section 4.2 and indeed, every set x satisfying the condition $0 \in x \wedge [\, (\forall y \in x)\, y \cup \{y\} \in x]$ must be in this formal sense infinite. A natural question occurs: Why is the axiom of infinity stated in precisely this way? Of course, there are many formulations which turn out to be equivalent. The existing formulation makes the development of natural numbers in Chapter 4 particularly smooth.

Historical debate: As there are no collections in sensory experience that are infinite, there was a considerable discussion, mostly predating the axiomatic development of set theory, regarding the use of infinite sets in mathematics.

Aristotle discerned between two kinds of infinity: the potential infinity and the actual infinity. A potentially infinite sequence is a sequence with a rule of extending it for an arbitrary number of steps (such as counting the natural numbers $0, 1, 2 \dots$). An actual infinity then grasps the whole result of repeating all these steps and views it as a completed object (the set of natural numbers). Zeno's paradoxes (5th century BC) have long been regarded as a proof that the actual infinity is an inherently contradictory concept. Bernard Bolzano, a Catholic philosopher, produced an argument that there are infinitely many distinct truths which must be all present in omniscient God's mind, and therefore, God's mind must be infinite (1851). This was intended as a defense of the use of infinite sets in mathematics. Poincaré and Hermann Weyl can be listed as important opponents of the use of infinite sets among 19th and 20th century mathematicians. *Finitism*, the rejection of the Axiom of Infinity, still has a small minority following among modern mathematicians. On a practical level, while a great deal of mathematics can be developed without the Axiom of Infinity, the formulations and proofs without the axiom become cumbersome and long.

The Axiom of Infinity states only that there is a set which contains every natural number and thus includes the desired set $\omega = \{0, 1, 2, \dots\}$ of natural numbers. In the following chapter, we will show how to obtain ω using the Axiom of Infinity together with the Axiom of Comprehension.

3.5 Axiom Schema of Comprehension

This axiom is also known as separation or collection. It is in fact an infinite collection of axioms, with one instance for each formula ϕ of set theory.

Definition 3.5.1. Let ϕ be a formula of set theory with $n + 1$ free variables for some natural number n. The instance of the *Axiom Schema of Comprehension* associated with ϕ is the following statement:

$$(\forall x)(\forall u_0)(\forall u_1) \ldots (\forall u_{n-1})(\exists y)(\forall z)[\, z \in y$$
$$\leftrightarrow [z \in x \,\wedge\, \phi(z, u_0, \ldots u_{n-1})]].$$

We use this axiom schema tacitly whenever we define sets using the *set builder notation*:

$$y = \{z \in x \colon \phi(x, u_0, u_1, \ldots u_n)\},$$

where x is called the *ambient* set. Comprehension makes it possible to form a great number of new sets. Given sets x, y, we can form the intersection $x \cap y = \{z \in x \colon z \in y\}$, the set difference $x \setminus y = \{z \in x \colon z \notin y\}$ as well as the symmetric difference $x \triangle y = (x \setminus y) \cup (y \setminus x)$. Given a nonempty set x of sets, we can form the intersection of all sets in x: $\bigcap x = \{z \in \bigcup x \colon (\forall y \in x)\, z \in y\}$.

We will use a similar process to form what we call classes.

Definition 3.5.2. A *class* is a collection C of sets such that there is a formula ϕ of $n + 1$ variables, and sets $u_0, \ldots u_{n-1}$, such that $z \in C \leftrightarrow \phi(z, u_0, u_1, \ldots, u_{n-1})$. A *proper class* is a class that is not a set.

The set builder notation: $C = \{z \colon \phi(z, u_0, u_1, \ldots u_{n-1})\}$ is often used to denote classes. A class may not be a set since the axiom schema of comprehension cannot be *a priori* applied due to the lack of the ambient set x. On some intuitive level, classes may fail to be sets on the account of being "too large".

Proposition 3.5.3. *For any nonempty class* C, $\bigcap C$ *is a set.*

Proof. Let $C = \{z : \phi(z, u_0, u_1, \ldots u_{n-1})\}$. Since C is nonempty, we may choose some set $A \in C$. Now, any element of $\bigcap C$ must also be an element of A. Thus,

$$\bigcap C = \{x \in A : (\forall z)\phi(z, u_0, u_1, \ldots u_{n-1}) \to x \in y\}. \qquad \square$$

This will enable us to prove the existence of the set ω of natural numbers in the following chapter.

Here is a standard example from algebra. We recall the following basic definitions.

Definition 3.5.4. A set G with binary operation $*$, inverse operation $^{-1}$, and identity e is a *group* if it satisfies the following:

- The operation $*$ is associative, that is, $a * (b * c) = (a * b) * c$ for all $a, b, c \in G$.
- $a * e = a = e * a$ for all $a \in G$.
- $a * a^{-1} = e = a^{-1} * a$ for all $a \in G$.

Definition 3.5.5. A nonempty subset H of a group G is a *subgroup* if for all $a, b \in H$, $a * b \in H$ and $a^{-1} \in H$ (it follows that $e \in H$).

Definition 3.5.6. For any subset A of a group G,

$$\langle A \rangle = \bigcap \{H : A \subseteq H \text{ and } H \text{ is a subgroup of } G\}.$$

For example, consider the group $\mathbb{Z}$ of integers with the addition operation and identity element 0. Then $\langle \{5\} \rangle = \{0, 5, -5, 10, -10, \ldots\}$ is the set of multiples of 5.

Since, for any subset A of a group G, G includes A and is a subgroup of itself, it follows from Proposition 3.5.3 that $\langle A \rangle$ is a set. In fact, it is not hard to see that $\langle A \rangle$ is a subgroup of G. The proof will be instructive.

Proposition 3.5.7. *For any subset A of a group G, $\langle A \rangle$ is a subgroup of G.*

Proof. Let $\mathcal{C}$ be the set of subgroups of G which include A, so that $\langle A \rangle = \bigcap \mathcal{C}$. Since the identity e belongs to every $H \in \mathcal{C}$, it follows that $e \in \langle A \rangle$, so that $\langle A \rangle$ is not empty.

Suppose that $a, b \in \langle A \rangle$ and let $H \in \mathcal{C}$. Then $a, b \in H$, so that $a * b \in H$ and $a^{-1} \in H$. Since this is true for all $H \in \mathcal{C}$, it follows that $a * b \in \langle A \rangle$ and $a^{-1} \in \langle A \rangle$. Thus, $\langle A \rangle$ is indeed a subgroup of G. □

Here is an example to show that not every class is a set.

Example 3.5.8. The class $\{z \colon z \notin z\}$ is not a set.

Proof. This is Russell's paradox. Suppose that there is a set x such that for every set z, $z \in x$ just in case $z \notin z$. Ask whether $x \in x$ or not. If $x \in x$, then by the definition of x, $x \notin x$, which is a contradiction. If $x \notin x$, then by the definition of x, $x \in x$ holds and we have a contradiction again. Both options lead to contradiction, proving that x does not exist. □

The universal class $\{z \colon z = z\}$ is often called V, the set theoretical universe. It is certainly not a set: If V were a set, then all classes would turn into sets by inserting the ambient set V into their definitions. However, we just produced a class which is not a set.

Historical debate: The formulation of the axiom schema of comprehension is motivated by the desire to avoid Russell's paradox. The use of the ambient set x makes it impossible to form sets such as $\{z \colon z \notin z\}$ since we are missing the ambient set: $y = \{z \in ? \colon z \notin z\}$. This trick circumvents all the known paradoxes, it comes naturally to all working mathematicians, and it does not present any extra difficulties in the development of mathematics in set theory.

There were other attempts to circumvent the paradoxes by limiting the syntactical nature of the formula ϕ used in the comprehension schema as opposed to requiring the existence of the ambient set x. One representative of these efforts is Quine's *New Foundations* (NF) axiom system [8]. Roughly stated, in NF, the formula ϕ has to be checked for circular use of $\in$ relation between its variables before it can be used to form a set. This allows the existence of the universal set $\{z \colon z = z\}$, but it also makes the development of natural numbers and general practical use extremely cumbersome. This seems like a very poor trade. As a result, NF is not used in mathematics today.

There was an objection to the possible use of *impredicative definitions* allowed by the present form of comprehension. Roughly stated, the objecting parties (including Russell and Poincaré) claimed that a set must not be defined by a formula which takes into account sets to which the defined set belongs (a formula ϕ defining a subset of some set x should not use the powerset of x as one of its parameters, for example). Such a definition would form, in their view, a *vicious circle*. It is challenging to make this objection precise. Mathematicians use impredicative definitions quite often and without care, for example, the usual proof of completeness of the real numbers contains a vicious circle in this view. Attempts to build mathematics without impredicative definitions turned out to be awkward. The school of thought objecting to impredicative definitions in mathematics mostly fizzled out before 1950.

The Axiom of Comprehension may be used to show the existence of various sets connected with relations and functions. Here is an example.

Definition 3.5.9. The Cartesian product $x \times y$ is the set of all ordered pairs (u, v) such that $u \in x$ and $v \in y$.

Proposition 3.5.10. *For any sets A and B, $A \times B$ is a set.*

Proof. First, we observe that the relation $x = \{a\}$ is definable by the formula $(\forall z)[z \in x \leftrightarrow z = a]$ and the relation $y = \{a, b\}$ is definable by the formula $(\forall z)[z \in y \leftrightarrow [z = a \lor z = b]]$. Then we can show (in the exercises) that the relation $x = (a, b)$ is also definable.

Now, let $a \in A$ and $b \in B$. Then both a and b are in $A \cup B$. Therefore, both $\{a\}$ and $\{a, b\}$ are in $\mathcal{P}(A \cup B)$. Hence, $(a, b) \in \mathcal{PP}(A \cup B)$. It follows that

$$A \times B = \{x \in \mathcal{PP}(A \cup B) : (\exists a \in A)(\exists b \in B)\, x = (a, b)\}.$$

Thus, $A \times B$ is a set by the Axiom of Comprehension. □

More generally, for a finite sequence $A_1, \ldots, A_n$ of sets, the product $A_1 \times \cdots \times A_n$ is also a set.

Exercises for Section 3.5

Exercise 3.5.1. Show that the intersection of a class and a set is a set.

Exercise 3.5.2. Show that $Dmn(R)$ and $Rng(R)$ are sets for any relation R.

Exercise 3.5.3. Show that for any relation $R \subseteq A \times B$, the inverse relation R^{-1} is also a set.

Exercise 3.5.4. Show that for any sets A and B, the set B^A of functions mapping A to B is a set.

Exercise 3.5.5. Define a bijection from $\mathcal{P}(A)$ to $\{0,1\}^A$.

Exercise 3.5.6. Show that for three sets A, B, C, the product $A \times B \times C$ is also a set.

3.6 Axiom of Choice

Definition 3.6.1. The *Axiom of Choice* (AC) is the following statement:

> *For every set x consisting of nonempty sets, there is a function f with* $\mathrm{Dmn}(f) = x$ *and* $(\forall y \in x)\ f(y) \in y$.

The function f is referred to as a *selector*.

Historical debate: The Axiom of Choice is the only axiom of set theory which asserts an existence of a set (the selector) without providing a formulaic description of that set. The Axiom of Infinity is presently stated in such a way as well, but (unlike AC) it can be reformulated to provide a definition of a certain infinite set. Naturally, AC provoked the most heated discussion of all the axioms.

Zermelo used AC in 1908 to show that the set of real numbers can be well ordered (see Section 5.2). This seemed counterintuitive, as the well-ordering of the reals is an extremely strong construction tool, and at the same time, it is entirely unclear how one could construct such a well-ordering. A number of people (including Lebesgue, Borel, and Russell) voiced various objections to AC as the main tool in Zermelo's theorem. A typical objection (Lebesgue) claimed that a proof of an existence of an object with a certain property, without a construction or definition of such an object, is not permissible. In the end, certain consequences of the axiom proved indispensable to the development of certain theories, such as Lebesgue's own theory of measure. The repeated implicit use of certain consequences of AC in the work of its very opponents also strengthened the case for adoption of the axiom.

One reason for the acceptance of the axiom was the lack of a constructive alternative. A plausible and useful alternative appeared in the 1960s in the form of Axiom of Determinacy (AD), asserting the existence of winning strategies in certain infinite two-player games [7]. At that point, the Axiom of Choice was already part of the orthodoxy, and so, AD remained on the sidelines.

Desirable consequences: The axiom of choice is helpful in the development of many mathematical theories. Typically, it allows proving general theorems about very large objects:

- (Algebra) Every vector space has a basis.
- (Dynamical systems) A continuous action of a compact semigroup has a fixed point.
- (Topology) The product of any family of compact spaces is compact.
- (Functional analysis) The Hahn–Banach theorem.

These applications of the Axiom of Choice are usually proved from certain equivalent formulations such as the Well-Ordering Principle and Zorn's Lemma. We will return to this topic later once we have discussed ordinals and transfinite recursion.

Undesirable consequences: Some weak consequences of AC are necessary for the development of theory of integration. However, its full form makes a completely harmonious integration theory impossible to achieve. It produces many "paradoxical" (a better word would be "counterintuitive") examples which force integration to apply to fairly regular functions and sets only.

- There is a nonintegrable function $f : [0, 1] \to [0, 1]$.
- (Banach–Tarski Paradox) There is a partition of the unit ball in $\mathbb{R}^3$ into several parts which can be reassembled by rigid motions to form two solid balls of unit radius.

The upshot: The axiom of choice is part of the mathematical orthodoxy today, and its suitability is not questioned or doubted by any significant number of mathematicians. A good mathematician notes its use though and (mostly) does not use it when an alternative proof without AC is available. The proof without AC will invariably yield more information than the AC proof. Almost every mathematical theorem asserting the existence of an object without (at least implicitly) providing its definition is a result of an application of the axiom of choice.

Definition 3.6.2. If x is a collection of nonempty sets, then $\prod x$, the *product* of x, is the collection of all selectors on x.

It is not difficult to see that $\prod x$ is a set. This is left as an exercise.

The Axiom of Choice asserts that the product of a collection of nonempty sets is nonempty. Recall that we have defined the product $\prod_{i \in I} A_i = \{f \colon I \to \bigcup_i A_i \colon (\forall i \in I) f(i) \in A_i\}$ of an indexed family of sets. An alternative version of the Axiom of Choice is to say that if $\{A_i : i \in I\}$ is an indexed family of nonempty sets, then the product $\prod_{i \in I} A_i$ is nonempty.

Proposition 3.6.3. *The following are equivalent:*

1. *The Axiom of Choice.*
2. *For any indexed family $\{A_i \colon i \in I\}$ of nonempty sets, $\prod_{i \in I} A_i \neq \emptyset$.*
3. *Zermelo's Principle: For any set A of nonempty, pairwise disjoint sets, there is a set B such that, for all $a \in A$, $B \cap a$ contains exactly one element.*
4. *The Relational Axiom of Choice: For any relation R, there is a function $F \subseteq R$ with $\mathrm{Dmn}(F) = \mathrm{Dmn}(R)$.*

Proof. Here is a proof that principles (1) and (2) are equivalent. The others equivalences are left as exercises.

Assume the Axiom of Choice and let $\{A_i : i \in I\}$ be an indexed family of nonempty sets. This means that there is a function f with domain I with $f(i) = A_i$. Now, let $A = \{A_i : i \in I\}$ and let g be a choice function for A. Then $g(A_i) \in A_i$ for each i. Putting these together, we see that $g(f(i)) \in A_i$ for each $i \in I$.

Next, assume the indexed version of choice and let A be any set of nonempty sets. Let $A = I$ and let $A_i = i$ define the indexed family $\{A_i : i \in I\}$. By the indexed version of choice, there is some function $h \in \prod_{i \in I} A_i$. This means that $h(i) \in A_i$ for each $i \in I$, and so, in this case $h(a) \in a$. Thus, A has a choice function. $\square$

Exercises for Section 3.6

Exercise 3.6.1. Show that for any set x, $\prod x$ is a set.

Exercise 3.6.2. Prove that Zermelo's Principle is equivalent to the Axiom of Choice.

Exercise 3.6.3. Prove that the Relational Axiom of Choice is equivalent to the Axiom of Choice.

3.7 Axiom Schema of Replacement

As was the case with the Axiom Schema of Comprehension, Replacement is not a single axiom but a schema including infinitely many axioms, one for each formula of set theory defining a class function. A formula $\phi(x,y)$, possibly with parameters, defines a (partial) class function F if for every x, there is at most one y such that $\phi(x,y)$, then $F(x) = y$ if and only if $\phi(x,y)$. To see the use of parameters, we can fix a set A and let $F(x) = x \cup A$. Here, we have $\phi(x,y) \iff (\forall z)z \in y \iff (z \in x \lor z \in A)$.

Definition 3.7.1. The *Axiom Schema of Replacement* states the following. If F is a class function and A is a set, then $F[A]$ is a set as well (this is often referred to as the "Axiom of Replacement").

   In standard mathematics, we frequently use a form of the set builder notation to write $\{F(x) : x \in A\}$ to define the image $F[A]$. For example, the set of squares of natural numbers may be written as $\{x^2 : x \in \mathbb{N}\}$ rather than $\{x \in \mathbb{N} : x \text{ is a square}\}$. This general usage is justified by the Axiom of Replacement. Replacement was a late contribution to the axiomatics of ZFC (1922). It is the only part of the axiomatics invented by Fraenkel. It is used almost exclusively for the internal needs of set theory; we will see that the development of ordinal numbers and well-orderings would be awkward without it. The only "mathematical" theorem for which it is known to be indispensable is the Borel determinacy theorem of Martin, ascertaining the existence of winning strategies in certain types of two-player infinite games [6].
   The Axiom of Comprehension follows from the Axiom of Replacement, but is kept in our system for convenience.

Theorem 3.7.2. *The Axiom of Replacement implies the Axiom of Comprehension.*

Proof. Assume the Axiom of Replacement. Let A be a set and let ϕ be a formula with parameters, and let $B = \{x \in A : \phi(x)\}$. Note that B is a class and we need to prove that it is in fact a set. If B is empty, then it is a set by the Empty Set Axiom. So, we may assume that $B \neq \emptyset$, and let $b \in B$. Now, define the function F so that $F(x) = x$, if $\phi(x)$, and $F(x) = b$, if $\neg\phi(x)$. Then F is a class function, since we have

$$y = F(x) \iff (\phi(x) \wedge y = x) \vee (\neg\phi(x) \wedge y = b).$$

It is easy to see that $F[A] = B$, so that B is a set by the Axiom of Replacement. Since A and ϕ were arbitrary, we see that the Axiom of Comprehension follows. $\square$

To obtain an infinite ordinal, we needed to introduce the Axiom of Infinity. However, the Axiom of Replacement will imply the existence of hierarchy of uncountable ordinals.

Exercises for Section 3.7

Exercise 3.7.1. Show that the Axiom Schema of Replacement is equivalent to the statement "Each class function with set domain is a set".

Exercise 3.7.2. The statement "the range of a set function is a set" can be proved without replacement. Use Comprehension to prove that, for any set function F and any set A, the image $F[A]$ of A under F is also a set.

Exercise 3.7.3. Show that there is no class injection from a proper class into a set.

3.8 Axiom of Regularity

The Axiom of Regularity is also known as Foundation or Well-foundedness.

Definition 3.8.1. The *Axiom of Regularity* states that

$$(\forall x)[x = 0 \vee (\exists y \in x)(\forall z \in x)\, z \notin y.$$

Restated, every nonempty set contains an $\in$-minimal element. This is the only axiom of set theory that explicitly limits the scope of the set-theoretic universe, ruling out the existence of sets such as the following.

Theorem 3.8.2. *Assuming the Axiom of Choice, the Axiom of Regularity is equivalent to the statement that there is no function f with domain ω such that $f(n+1) \in f(n)$ for all n.*

Proof. Suppose first that there is a function f such that $f(n+1) \in f(n)$ for all n. Then the range $f[\omega]$ has no minimal element, which contradicts Regularity.

For the other direction, suppose that there is a nonempty set S which has no minimal element. Let $f(0)$ be some element of S. Since S has no minimal element, it follows that, for any $a \in S$, $S_a = \{x \in S : x \in a\}$ is nonempty. Let Φ be a choice function so that $\Phi(a) \in S_a$ for each $a \in S$; that is, $\Phi(a) \in a$. Now, recursively define f so that $f(n+1) = \Phi(f(n))$ for each n. Then f is the desired function such that $f(n+1) \in f(n)$ for all n. $\square$

We will discuss in detail the idea of recursive definitions in the following chapter.

The motivation behind the adoption of this axiom lies in the fact that the development of common mathematical notions within set theory uses sets that always, and of necessity, satisfy regularity. The formal development of set theory is smoother with the axiom as well. The present form of the axiom is due to von Neumann [11]. Mathematical interest in the phenomena arising when the Axiom of Regularity is denied has been marginal [1].

Exercise for Section 3.8

Exercise 3.8.1. Show that for any sets x and y with $x \in y$, x is an $\in$-minimal element of y if and only if $x \cap y = \emptyset$.

Exercise 3.8.2. Use the Axiom of Regularity to show that there is no set x with $x \in x$, and there are no sets x, y such that $x \in y \in x$.

Chapter 4

Natural Numbers and Countable Sets

4.1 Von Neumann's Natural Numbers

The purpose of this section is to develop the natural numbers in ZFC.

Definition 4.1.1. For a set x, write $s(x) = x \cup \{x\}$. A set y is *inductive* if $0 \in y$ and for all x, $x \in y$ implies $s(x) \in y$.

The Axiom of Infinity says precisely that there is an inductive set.

Definition 4.1.2 (von Neumann). The $\subseteq$-smallest inductive set is denoted by $\mathbb{N}$, or ω. A set x is a *natural number* if $x \in \omega$. For natural numbers x, $s(x)$ is the *successor* of x.

The set ω is intended to be the set $\{0, 1, 2, \ldots\}$ of natural numbers, where we let $1 = s(0)$, $2 = s(s(0))$, and so on. With every definition of this sort, one has to make sure that it actually makes sense. This is the contents of the following theorem.

Theorem 4.1.3. *There is an $\subseteq$-smallest inductive set.*

Proof. Let ω be the intersection of all inductive sets. Clearly, if there is an $\subseteq$-smallest inductive set, then ω must be it, so it is enough to verify that ω is inductive.

First, we verify that ω is in fact a set. Let $\mathcal{C}$ be the class of all inductive sets. The class $\mathcal{C}$ is non-empty by the Axiom of Infinity. Thus, $\bigcap \mathcal{C} = \omega$ is a set by Proposition 3.5.3.

Second, we check that ω is itself an inductive set. For that, we have to verify that $0 \in \omega$ and for every $x \in \omega$, $s(x) \in \omega$ holds. As 0 belongs to every inductive set, $0 \in \omega$ by the definition of ω. Now, suppose that $x \in \omega$; we must show that $s(x) \in \omega$. For every inductive set y, $x \in y$ holds by the definition of ω. As y is inductive, $s(x) \in y$ as well. We have just proved that $s(x)$ belongs to every inductive set, in other words, $s(x) \in \omega$. This completes the proof. □

The main feature of ω is that we can use induction to prove various statements about natural numbers. This is the content of the following theorem.

Theorem 4.1.4 (Induction). *Suppose that ϕ is a formula, $\phi(0)$ holds, and $(\forall x \in \omega)[\phi(x) \to \phi(s(x))]$ also holds. Then $(\forall x \in \omega)\,\phi(x)$ holds.*

Proof. Consider the set $y = \{x \in \omega : \phi(x)\}$. We show that y is an inductive set. Then, since ω is the smallest inductive set, it follows that $y = \omega$, in other words, $(\forall x \in \omega)\,\phi(x)$ as desired.

Indeed, $0 \in y$ as $\phi(0)$ holds. If $x \in y$, then $s(x) \in y$ as well by the assumptions on the formula ϕ. It follows that y is an inductive set as desired. □

We use the standard terminology for induction: $\phi(0)$ is the *base step*, the implication $\phi(x) \to \phi(s(x))$ is the *induction step*, and the formula $\phi(x)$ in the induction step is the *induction hypothesis*. The next step is to verify that $\in$ on ω is a strict linear ordering that emulates the properties of natural numbers.

Theorem 4.1.5 (Linear Ordering of Natural Numbers).

1. *If $x \in \omega$ and $y \in x$, then $y \in \omega$.*
2. *The relation $\in$ is a strict linear ordering on ω.*

Proof. For (1), let $\phi(x)$ be the statement $(\forall y \in x)\,y \in \omega$. We prove $(\forall x \in \omega)\,\phi(x)$ by induction on x.

Base Step. The statement $\phi(0)$ holds as its first universal quantifier ranges over the empty set.

Induction Step. Suppose that $\phi(x)$ holds. To prove $\phi(s(x))$, let $y \in s(x)$. *Either* $y \in x$, in which case $y \in \omega$ by the induction hypothesis. *Or* $y = x$, in which case $y \in \omega$ since $x \in \omega$. This proves (1).

To prove (2), we have to verify the transitivity and linearity of $\in$ on ω. We start with transitivity. Let $\phi(x)$ be the statement $(\forall y \in x)(\forall z \in y)\, z \in x$. We prove $(\forall x \in \omega)\, \phi(x)$ by induction on $x \in \omega$.

Base Step. The statement $\phi(0)$ holds as its first universal quantifier ranges over the empty set.

Induction Step. Suppose that $\phi(x)$ holds; we work to verify $\phi(s(x))$. Let $y \in s(x)$ and $z \in y$. By the definition of $s(x)$, there are two cases to consider. *Either* $y \in x$ so that by the induction hypothesis $z \in x$, and as $x \subseteq s(x)$, $z \in s(x)$ holds. *Or* $y = x$ so that $z \in x$ and as $x \subseteq s(x)$, $z \in s(x)$ holds again. This confirms the induction step and proves transitivity.

Next, we proceed to linearity. The following two preliminary claims will be useful:

Claim 4.1.6. *For every $y \in \omega$, $0 = y$ or $0 \in y$.*

Proof. Let $\psi(y)$ be the statement $0 = y \ \vee \ 0 \in y$; we prove $(\forall y \in \omega)\, \psi(y)$ by induction on $y \in \omega$.

Base Step. $\psi(0)$ holds as $y = 0$ is one of the disjuncts.

Induction Step. Suppose that $\psi(y)$ holds; we verify $\psi(s(y))$. The induction hypothesis offers two cases. *Either* $y = 0$, in which case $y = 0 \in s(y)$ by the definition of $s(y)$. *Or* $0 \in y$ so that $0 \in s(y)$, since $y \subseteq s(y)$. In both cases, the induction step has been confirmed. $\square$

Claim 4.1.7. *For every $y \in \omega$, for every $x \in y$, $s(x) \in s(y)$ holds.*

Proof. Let $\psi(y)$ be the statement $(\forall x \in y)\, s(x) \in s(y)$; by induction on $y \in \omega$, we prove $(\forall y \in \omega)\, \psi(y)$.

Base Step. $\psi(0)$ is trivially true as its universal quantifier ranges over the empty set.

Induction Step. Assume that $\psi(y)$ holds; we work to verify $\psi(s(y))$. Let $x \in s(y)$ be any element. By the definition of $s(y)$, there are two cases to consider. *Either* $x \in y$ so that by the induction hypothesis $s(x) \in s(y)$, and as $s(y) \subseteq s(s(y))$, $s(x) \in s(s(y))$ holds. *Or* $x = y$, in which case $s(x) = s(y) \in s(s(y))$ by the definition of $s(s(y))$. In both cases, the induction step has been confirmed. $\square$

Now, let $\phi(x)$ be the statement $(\forall y \in \omega)[\, x = y \,\vee\, x \in y \,\vee\, y \in x\,]$. By induction on $x \in \omega$, we prove $(\forall x \in \omega)\,\phi(x)$.

Base Step. The statement $\phi(0)$ follows from Claim 4.1.6.

Induction Step. Suppose that $\phi(x)$ holds; we work to verify $\phi(s(x))$. Let $y \in \omega$ be arbitrary. The induction hypothesis yields a split into three cases. *Either* $y \in x$, and thus, as $x \subseteq s(x)$, $y \in s(x)$. *Or*, $y = x$ and then $y \in s(x)$ by the definition of $s(x)$. *Or*, $x \in y$, and then by Claim 4.1.7, $s(x) \in s(y)$ holds, which by the definition of $s(y)$ says that either $s(x) \in y$ or $s(x) = y$. In all cases, the induction step has been confirmed.

The linearity of the $\in$ ordering on natural numbers has been verified.

$\square$

For natural numbers m and n, let $m < n$ if and only if $m \in n$. Thus, as a set, every natural number is exactly the set of all natural numbers smaller than it. Now, a rather routine induction argument (see the exercises) shows that for every $x \in \omega$, $s(x)$ is the smallest natural number larger than x, and for every non-zero natural number x, there is a largest number y smaller than x such that $x = s(y)$. We sometimes write $n + 1$ for $s(n)$, since when addition of natural numbers is defined in Section 4.3, it is seen that $s(n) = n + 1$.

It is important to note that our smallest inductive set ω is in fact just the set $\{0, 1, 2, \dots\}$ of natural numbers.

There is a stronger form of induction, known as *course-of-values* induction, where the induction step assumes that $\phi(y)$ holds for all $y < x$.

Example 4.1.8. The theorem that every natural number $n > 1$ has a prime factor is proved by course-of-values induction. Suppose that for all y with $1 < y < x$, y has a prime factor and suppose, by way of contradiction, that x has no prime factor. Then x itself is not prime, so we must have $x = y \cdot z$ with $1 < y, z < x$. But the induction hypothesis tells us that y (as well as z) has a prime factor p and then p is also a prime factor of x.

Exercises for Section 4.1

Exercise 4.1.1. Write down the natural number 4 as a set using the set builder notation and the symbol 0. Count the number of symbols. (For example, 3 can be written as $\{0, \{0\}, \{0, \{0\}\}\}$ having 15 symbols, counting the commas, set brackets, and 0's). Can you find a general formula for the number of symbols need to write the number n?

Exercise 4.1.2. By induction on $x \in \omega$, show that if $x \neq 0$ is a natural number then it has a predecessor, i.e., a number y which is largest among all numbers smaller than x, and such that $x = s(y)$.

Exercise 4.1.3. Show that course-of-values induction is valid. That is, suppose ϕ is a formula such that $\phi(x)$ is true whenever $\phi(y)$ is true for all $y < x$. Then show that $\{x : \phi(x)\}$ is an inductive set.

Exercise 4.1.4. Without using the Axiom of Regularity, show that every nonempty subset of ω has an $\in$-smallest element.

Exercise 4.1.5. Show that, for two natural numbers m and n, $m \subseteq n$ if and only if $m \leq n$.

Exercise 4.1.6. Show that $\bigcup s(n) = n$ for each $n \in \omega$.

4.2 Finite and Infinite Sets

The purpose of this section is to develop the notion of finiteness for sets. One reasonable way to proceed is to define a set to be finite if it is in a bijection with some natural number. We use a different definition which has the virtues of being more intellectually stimulating, very efficient in proofs, and independent of the development of ω:

Definition 4.2.1 (Tarski). A set x is *finite* if every non-empty set $a \subseteq \mathcal{P}(x)$ has a $\subseteq$-minimal element, i.e., a set $y \in a$ such that no $z \in a$ is a proper subset of y. A set is *infinite* if it is not finite.

Note that the existence of a minimal set here is equivalent to the existence of a maximal set, since a $\subseteq$-maximal set is a minimal set in the family of complements of the sets in x.

With a somewhat slick definition of this sort, it is necessary to verify that it corresponds to the intuitive notion of finiteness. We first provide a basic example of a finite and infinite set:

Example 4.2.2. 0 is a finite set and ω is an infinite set.

Proof. $\mathcal{P}(0) = \{0\}$ so that, if $a \subseteq \mathcal{P}(0)$ is a non-empty set, then $a = \{0\}$, so 0 is a $\subseteq$-minimal of a.

To see that ω is infinite, for every $n \in \omega$, let $y_n = \{m \in \omega : n \in m\}$ and let $a = \{y_n : n \in \omega\}$. This is a subset of $\mathcal{P}(\omega)$; let us show that it has no $\subseteq$-minimal element. Suppose y_n is such a minimal element. Then $y_{n+1} \in a$ is a proper subset of y_n, contradicting the minimality of y_n. $\qquad\qquad\square$

Theorem 4.2.3. *The class of finite sets is closed under the following operations:*

1. *taking a subset;*
2. *adding a single element to a set;*
3. *taking unions;*
4. *forming the image under a surjection;*
5. *taking the powerset.*

Proof. For (1), suppose that x is finite and $y \subseteq x$; we must argue that y is finite. Let $a \subseteq \mathcal{P}(y)$ be a non-empty set; we must show that a has a $\subseteq$-minimal element. Since $y \subseteq x$, it is the case that $a \subseteq \mathcal{P}(x)$. As x is finite, a must have a $\subseteq$-minimal element as desired.

For (2), suppose that x is finite and i is any set; we must verify that the set $y = x \cup \{i\}$ is finite. Let $a \subseteq \mathcal{P}(y)$ be a non-empty set; we must produce a $\subseteq$-minimal element of a. Let $b = \{u \cap x : u \in a\}$. This is a non-empty subset of $\mathcal{P}(x)$; as x is assumed to be finite, the set b has a $\subseteq$-minimal element v. There are now two cases to consider. Either $v \in a$, in which case v is a $\subseteq$-minimal element of a. Or, $v \notin a$, in which case $u = v \cup \{i\}$ is a $\subseteq$-minimal element of a. This completes the proof of (2).

For (3), assume for contradiction that x, y are finite and $x \cup y$ is not. Let $a = \{z \subseteq x : z \cup y \text{ is not finite}\}$. This is a non-empty subset of x containing at least x as an element. Since x is finite, the set a has a $\subseteq$-minimal element, say u. The set u must be non-empty since $y \cup 0 = y$ is a finite set. Let $i \in u$ be an arbitrary element, and let $v = u \setminus \{i\}$. By the minimality of u, $y \cup v$ is finite. By (2), $y \cup v \cup \{i\}$ is finite as well. As $y \cup v \cup \{i\} = y \cup u$, this contradicts the assumption that $u \in a$.

For (4), assume for contradiction that x is a finite set, $f : x \to y$ is a surjection, and y is not finite. Let $a = \{z \subseteq x : f[z] \text{ is not}$

finite}. This is a subset of $\mathcal{P}(x)$ which by our contradictory assumption contains x as an element and therefore a is non-empty. Note that $u \neq 0$, since $f(0) = 0$ is finite by Example 4.2.2. As x is finite, a contains a $\subseteq$-minimal element u. Let $i \in u$ be an arbitrary element, and let $v = u \setminus \{i\}$. Then $v \notin a$ by the minimal choice of u, and so $f[v]$ is finite. However, $f[u] = f[v] \cup \{f(i)\}$, which is finite by (2), contradicting the assumption that $u \in a$.

For (5), assume for contradiction that x is finite and $\mathcal{P}(x)$ is not finite. Let $a = \{y \subseteq x : \mathcal{P}(y) \text{ is not finite}\}$. This is a non-empty set, containing at least x as an element. Let u be a $\subseteq$-minimal element of a. Note that $u \neq 0$, since $\mathcal{P}0) = \{0\}$ is finite by part (2). Pick an element $i \in u$ and consider the set $v = u \setminus \{i\}$. Then, $\mathcal{P}(u) = \mathcal{P}(v) \cup \{z \cup \{i\} : z \in \mathcal{P}(v)\}$. The first set in the union is finite by the minimality of u, and the second is a surjective image of the first, therefore finite as well. By the previous items, $\mathcal{P}(u)$ is finite, and this is a contradiction to the assumption that $u \in a$. $\square$

Here are two corollaries.

Corollary 4.2.4. *Every natural number is finite.*

Proof. This is proven by induction on n.

Base Step. 0 is finite by Example 4.2.2.

Induction Step. Suppose by induction that n is finite. Then $n + 1 = n \cup \{n\}$ is finite by part (2) of Theorem 4.2.3. $\square$

Corollary 4.2.5. *An injective image of an infinite set is infinite.*

Proof. Let x be an infinite set and let f be an injection with $\mathrm{Dmn}(f) = x$. We have to argue that $y = Rng(f)$ is infinite. Note first that $g = f^{-1}$ is a surjective function from y to x. So, if y were finite, then $g[y] = x$ would be finite as well by Theorem 4.2.3(4), contradicting the assumption that x is infinite. $\square$

Here is a very useful result.

Theorem 4.2.6. *A finite union of finite sets is finite.*

Proof. Suppose that a set A is finite and each of its elements a is also finite. Suppose by way of contradiction that $\bigcup A$ is not finite and consider $C = \{x \subseteq A : \bigcup x \text{ is not finite}\}$. By assumption, $A \in C$ so that C is non-empty. Since A is finite, C must contain a $\subseteq$-minimal element B. $B \neq \emptyset$ since $\bigcup \emptyset = \emptyset$ is finite. So, let $a \in B$. Since B is minimal in C, $B \setminus \{a\} \notin C$, so $\bigcup(B \setminus \{a\})$ is finite. Then $\bigcup B = a \cup \bigcup(B \setminus \{a\})$ is finite by part (3) of Theorem 4.2.3. This contradicts our assumption that $B \in C$. $\qquad\square$

The following principle is used throughout mathematics. It is a contrapositive of Theorem 4.2.6.

Corollary 4.2.7 (Pigeonhole Principle). *If an infinite set A is partitioned into finitely many subsets, then at least one of those subsets is infinite.*

The final theorems of this section characterize finiteness in terms of natural numbers. This allows one to prove theorems about finite sets by induction on their size. Note that the treatment of finiteness up to this point has not used natural numbers at all.

Theorem 4.2.8. *A set A is finite if and only if there is a bijection between A and some natural number.*

Proof. First, suppose that there is a bijection between A and a natural number n. If A were infinite, then n would be infinite by Corollary 4.2.5. But this contradicts Corollary 4.2.4. Hence, A must be finite.

Next, suppose that A is finite and suppose by way of contradiction that A is not bijective with any natural number. Let

$$B = \{x \subseteq A : x \text{ is not bijective with any } n \in \mathbb{N}\}.$$

Then $A \in B$ so that B is not empty and hence has a $\subseteq$-minimal element C. Then $C \neq \emptyset$, since $\emptyset = 0$. Thus, C has a member a. By assumption, $C \setminus \{a\}$ is bijective with some natural number n. Let $f : n \to C \setminus \{a\}$ be a bijection. Now, define $g : n + 1 \to C$ by $g(i) = f(i)$ if $i < n$ and $g(n) = a$. This is clearly a bijection, contradicting our assumption. $\qquad\square$

Theorem 4.2.9. *Let A be a set. The following are equivalent for any set A:*

1. *A is infinite.*
2. *A contains an injective image of ω.*
3. *There is a bijection between A and a proper subset B of A.*

The proof uses the Axiom of Choice and, in fact, cannot be proved without the Axiom of Choice. The proof also uses the ability to define functions by recursion, which is justified in the following section.

Proof. (1) $\implies$ (2): Let A be an infinite set; we must produce an injection from ω to A. Use the Axiom of Choice to produce a selector function $H \colon \mathcal{P}(A) \setminus \{0\} \to A$. Now, consider the recursive definition of a function $F \colon \omega \to A$ given by $F(n) = H(A \setminus F[n])$. Note that for every natural number $n \in \omega$, the set $F[n] \subseteq A$ is finite by Theorem 4.2.3. Since A is infinite, the set $A \setminus F[n]$ is non-empty, so the value $F(n) = H(x \setminus F[n])$ is defined and different from all values of $F(m)$ for $m \in n$. It is then clear that F is the desired injection from ω to A.

(2) $\implies$ (3): Let $F \colon \omega \to A$ be an injection, let $C = F[\omega]$, and let $B = A \setminus \{F(0)\}$. Then we can define a bijection G from B to A by letting $G(F(n+1)) = F(n)$ for $x \in C \setminus \{F(0)\}$ and letting $G(x) = x$ for $x \notin C$.

(3) $\implies$ (1): Let H be a bijection from A to a proper subset B of A and let $b \in A \setminus B$. Since H is one-to-one, it follows that the elements $b_0 = b, b_1 = H(b), b_2 = H(H(b)), \ldots$ are distinct; note that $b \neq H(x)$ for any x since $b \notin A$. Now, this gives us a family $\{b_0\} \subseteq \{b_0, b_1\} \subseteq \cdots$ which has no maximal element. Hence, A is infinite. $\qquad\square$

Exercises for Section 4.2

In the following exercises, use Tarski's definition of finiteness:

Exercise 4.2.1. Let x be a finite non-empty set and $\leq$ a linear ordering on x. Prove that x has a largest element in the sense of the ordering $\leq$.

Exercise 4.2.2. Prove that the product of two finite sets is finite.

Exercise 4.2.3. Prove without the Axiom of Choice that if x is a finite set consisting of non-empty sets, then x has a selector.

Exercise 4.2.4. Without the use of Axiom of Infinity show that the existence of an infinite set is equivalent to the existence of an inductive set.

Exercise 4.2.5. Show that if B is a finite set and $B \subseteq \bigcup_{n \in \mathbb{N}} A_n$, then there exists $n \in \mathbb{N}$ such that $B \subseteq \bigcup_{i \leq n} A_i$.

4.3 Inductive and Recursive Definability

Now, that we have the set ω of natural numbers; we may define subsets of ω and functions on ω by a general form of inductive definability.

Definition 4.3.1. A *monotone operator* over a set A is a function $\Gamma : \mathcal{P}(A) \to \mathcal{P}(A)$ such that, for any $X, Y \in \mathcal{P}(A)$, $X \subseteq Y$ implies $\Gamma(X) \subseteq \Gamma(Y)$:

1. A set $X \in \mathcal{P}(A)$ is said to be a *fixed point* of Γ if $\Gamma(X) = X$.
2. Γ is said to be *finitary* if for any set $X \subseteq A$ and any $a \in A$, if $a \in \Gamma(X)$, then there is a finite subset S of X such that $a \in \Gamma(S)$.
3. A fixed point X of Γ is the *least fixed point* of Γ if $X \subseteq Y$ for all fixed points Y of Γ. The least fixed point of Γ is the set inductively defined by Γ.

We show in Theorem 4.3.3 that any monotone operator over a set A has a least fixed point. For example, consider the monotone operator on ω defined by $\Gamma(X) = \{0\} \cup \{s(x) : x \in X\}$. If $\Gamma(X) = X$, then clearly X contains 0 and is closed under the successor function. Thus, the fixed points of Γ are exactly what were called "inductive" sets in Section 4.1. The definition of ω thus makes it the least fixed point of Γ. This operator is finitary since if $n \in \Gamma(X)$, then either $n = 0$, so that $n \in \Gamma(\emptyset)$, or $n = s(x)$ for some $x \in X$, so that $n \in \Gamma(\{x\})$. Here is a standard example from logic.

Example 4.3.2. Let A be the set of finite strings over the alphabet $\{\neg, \vee, p, q, (,)\}$ and let

$$\Gamma(X) = X \cup \{p, q\} \cup \{\neg\phi : \phi \in X\} \cup \{(\phi \vee \psi) : \phi, \psi \in X\}.$$

Then Γ is a finitary monotone operator. The least fixed point of Γ is the set of all propositional sentences in the variables p, q — in the restricted language with connectives $\neg, \vee$.

We also consider *class operators* Γ where there is a formula ϕ such that, for all sets X and Y, $\Gamma(X) = Y$ if and only if $\phi(X, Y)$.

The notions of finitary operators and fixed points are similarly defined for class operators. The class operator $\bigcup$ is monotone and also finitary. The operator $\mathcal{P}$ is monotone but in general is not finitary.

Theorem 4.3.3. *For any monotone operator* Γ *on a set* A, *there exists a least fixed point* M.

Proof. Let $\mathcal{C} = \{X \in \mathcal{P}(A) : \Gamma(X) \subseteq X\}$. $\mathcal{C}$ is non-empty since $A \in \mathcal{C}$. Let $M = \bigcap \mathcal{C}$. Then immediately we have that $M \subseteq X$ for any $X \in \mathcal{C}$. By the definition of monotonicity, it follows that $\Gamma(M) \subseteq \Gamma(X)$ for any $X \in \mathcal{C}$. Thus, $\Gamma(M) \subseteq M$, and then by monotonicity, $\Gamma(\Gamma(M)) \subseteq \Gamma(M)$. This means that $\Gamma(M) \in \mathcal{C}$, which implies that $M \subseteq \Gamma(M)$. So, M is indeed a fixed point and is a subset of every other fixed point. $\qquad\square$

There is another approach to the fixed point. Here we examine this only for finitary operators.

Definition 4.3.4. For any monotone operator Γ on a set A, let $\Gamma^0 = \emptyset$, let $\Gamma^{n+1} = \Gamma(\Gamma^n)$, and let $\Gamma^\omega = \bigcup_{n \in \omega} \Gamma^n$.

It is easy to see by induction that Γ^n exists for each n. We see in the following that the union Γ^ω is a set, as is the least fixed point of the operator Γ. Here is another way to prove this. The following lemma is needed.

Proposition 4.3.5. *For any monotone operator* Γ *on a set* A, $\{(n, x) : x \in \Gamma^n\}$ *is a set and* Γ^ω *is a set.*

Proof. Define the monotone operator Δ on the set $\omega \times A$ by

$$(n + 1, x) \in \Delta(Y) \iff x \in \Gamma(\{y : (n, y) \in Y\}).$$

Note first that $(0, x) \notin \Delta(Y)$ for any x and Y. To see that Δ is monotone, suppose that $Y \subseteq Z \subseteq \omega \times A$ and that $(n + 1, x) \in \Delta(Y)$. Then $\{y : (n, y) \in Y\} \subseteq \{y : (n, y) \in Z\}$. Since Γ is monotone, it follows that $(n + 1, x) \in \Delta(Z)$.

Let $B = \{(n, x) : x \in \Gamma^n\}$, and let C be the least fixed point of Δ. We claim that $C = B$.

First, observe that $\Delta(B) = B$. That is, $(n + 1, x) \in \Delta(B)$ if and only if $x \in \Gamma(\{y : (n, y) \in B\})$, which holds if and only if $x \in \Gamma(\Gamma^n)$, which holds if and only if $x \in \Gamma^{n+1}$. It follows that $C \subseteq B$.

Next, suppose that D is any set with $\Delta(D) \subseteq D$. We show by induction on n that $\Gamma^n \subseteq \{x : (n, x) \in D\}$ so that $B \subseteq D$.

Base Step. For $n = 0$, this follows since $\Gamma^0 = \emptyset$.

Induction Step. Suppose that $\Gamma^n \subseteq \{x : (n, x) \in D\}$ and let $y \in \Gamma^{n+1} = \Gamma(\Gamma^n)$. Then by monotonicity, $y \in \Gamma(\{x : (n, x) \in D\})$ so that $(n + 1, y) \in \Delta(D)$ and by assumption, $(n + 1, y) \in D$. Thus, $y \in \{x : (n + 1, x) \in D\}$.

Finally, we have $x \in \Gamma^\omega \iff (\exists n)(n, x) \in C$. Thus, Γ^ω is a set since C is a set by Theorem 4.3.3. $\qquad\square$

Example 4.3.6. The set of even numbers may be defined by the operator $\Gamma(X) = \{0\} \cup \{s(s(x)) : x \in X\}$. Here we have $\Gamma^n = \{2i : i < n\}$.

In Example 4.3.2, $\Gamma^1 = \{p, q, \neg p, \neg q, (p \vee q)\}$. Here is a standard example from group theory.

Example 4.3.7. Given a group G with multiplication $*$, identity e, and inverse $^{-1}$ and a subset A of G, the subgroup $\langle A \rangle$ generated by A may be defined as the least fixed point of the finitary monotone operator Γ_A, where

$$\Gamma_A(X) = A \cup \{e, x * y, x^{-1} : x, y \in X\}.$$

That is, $z \in \Gamma_A(X)$ if and only if at least one of the following is true:

1. $z \in A$;
2. $z = e$;
3. $z^{-1} \in X$;
4. $z = x * y$ for some $x, y \in X$.

For the multiplicative group $\mathbb{Z}$ of integers, consider the set $A = \{2, 5\}$. Then we have

$$\Gamma^1 = \{1, 2, 4, 5, \tfrac{1}{2}, \tfrac{1}{5}, 10, 25\},$$
$$\Gamma^2 = \{1, 2, 4, 5, 8, 10, 20, 25, 40, 50, 100, 125, 250, 625, \tfrac{2}{5}, \tfrac{5}{2}, \tfrac{4}{5}, \tfrac{25}{2}, \tfrac{1}{4}, \tfrac{1}{10}, \tfrac{1}{25}\},$$

and so on. The subgroup $\langle A \rangle$ generated by A thus consists of all rationals of the form $2^i 5^j$, where i and j are integers.

Theorem 4.3.8. *For any finitary monotone operator on a set A, Γ^ω is the least fixed point of Γ.*

Proof. Let M be the least fixed point of Γ. First, we show by a simple induction that each $\Gamma^n \subseteq M$ so that $\Gamma^\omega \subseteq M$. Certainly, $\Gamma^0 = \emptyset \subseteq M$. Now, suppose that $\Gamma^n \subseteq M$. Then by monotonicity, $\Gamma^{n+1} \subseteq \Gamma(M) = M$.

Next, we show that Γ^ω is a fixed point of Γ. Suppose that $a \in \Gamma(\Gamma^\omega)$. Since Γ is finitary, there is a finite set $S \subseteq \Gamma^\omega$ such that $a \in \Gamma(S)$. Since S is finite and $\Gamma^\omega = \bigcup_n \Gamma^n$, it follows from Exercise 4.2.5 that there is some finite m such that $S \subseteq \Gamma^m$. Since Γ is monotone, $a \in \Gamma(\Gamma^m) = \Gamma^{m+1}$ and therefore $a \in \Gamma^\omega$. $\qquad\square$

Functions on a fixed set may also be defined inductively.

Example 4.3.9. The graph of the addition function on ω is $A = \{(x, y, z) \in \omega^3 : x + y = z\}$. Addition may be defined recursively by having $x + 0 = x$ for all x and $x + (s(y)) = s(x + y)$ for all x and y. Then A is the least fixed point of the operator Γ, where

$$\Gamma(X) = \{(x, 0, x) : x \in \omega\} \cup \{(x, s(y), s(z)) : (x, y, z) \in X\}.$$

Given the recursive definition of addition, we can now derive various properties of addition.

Proposition 4.3.10. *For all $n \in \mathbb{N}$, $0 + n = n$.*

Proof. Since the graph of the addition function is the least fixed point of the operator Γ given above, it suffices to show that for any set S, if $\Gamma(S) \subseteq S$, then $(0, n, n) \in S$ for all n. Given such a set S, let $X = \{n : (0, n, n) \in S\}$. We know by the definition of Γ that $(0, 0, 0) \in \Gamma(S)$ and therefore $(0, 0, 0) \in S$ and $0 \in X$. Now, suppose that $n \in X$. Then $(0, n, n) \in S$. Then $(0, s(n), s(n)) \in \Gamma(S)$, so $(0, s(n), s(n)) \in S$ and therefore $s(n) \in X$. It follows from the definition of ω as the least inductive set that $X = \omega$. That is, $0+n = n$ for all $n \in \omega$. $\qquad\square$

Just as induction proofs sometimes require a course-of-values induction, a recursive definition of the function F may need more than one previous value from $F(0), \ldots, F(n)$ to compute $F(n+1)$.

Example 4.3.11. The classic Fibonacci sequence is defined by the function F, where $F(0) = 0$, $F(1) = 1$, and $F(n + 2) =$

$F(n) + F(n + 1)$. Then the graph of the Fibonacci function is the smallest set F containing $(0, 0)$ and $(1, 1)$ such that whenever (n, x) and $(n + 1, y)$ are both in F, then $(n + 1, x + y)$ is in F.

Now, suppose we want to prove that $F(n) \geq 1.5^{n-2}$ for $n \geq 2$. The base case is $n = 2$, where $F(2) = 1 = 1.5^0$. For the induction step, we assume that $F(n) \geq 1.5^{n-2}$ and $F(n + 1) \geq 1.5^{n-1}$. Then $F(n + 2) = F(n) + F(n + 1) \geq 1.5^{n-2} + 1.5^{n-1} = 1.5^{n-2}(1 + 1.5) \geq 1.5^{n-2}1.5^2 = 1.5^n$, as desired.

We need to consider a general way to use the values $G(0), \ldots, G(n - 1)$ to compute $G(n)$.

Definition 4.3.12. For any set A and any function G (including a class function), let $G \restriction A$ be the restriction of G to domain A. That is, we identify G with its graph and let $G \restriction A = \{(x, y) \in G : x \in A\}$. In particular, if $Dmn(G) = \omega$, then $G \restriction n$ denotes the restriction of G to $n = \{0, 1, \ldots, n - 1\}$.

Thus far, these notions of inductive and recursive definability of functions have required that the domain and range of the function be at least included in some given set.

Next, we consider the more complicated case when we have a class function defined on ω. Here we are thinking of iterating the Union operator as well as the Power Set operator. The problem here is that there is not necessarily any universal set A such that $\Gamma(A) \subseteq A$ so that the least fixed point cannot be defined without further effort.

Theorem 4.3.13. *Suppose that F is a class function such that $F(n, x)$ is defined for every $n \in \omega$ and every set x. Let the set A_0 be given. Then there is a unique class function G such that $Dmn(G) = \omega$, $G(0) = A_0$, and for every $n \in \omega$, $G(n + 1) = F(n, G(n))$. Therefore, the image $G[\omega]$ is a set.*

Proof. The intuitive definition of G as a class function is that $G(n) = y$ if and only if there is a sequence $G(0) = A_0$, $G(1) = F(0, G(0))$, $\ldots$, $G(n) = F(n - 1, G(n - 1))$. This is formalized as follows:

We show by induction that for each n,

1. $G(n)$ is a set;
2. $G \restriction n$ is a set.

Base Step. For $n = 0$, we have $G(0) = A_0$ is a set and we have $G \restriction n = \emptyset$.

Induction Step. Suppose that $G(0), \ldots, G(n)$ are sets and that $G \restriction n$ is a set. Then $G(n + 1) = F(n, G(n))$ is a set since F is a class function and $G \restriction n + 1 = G \restriction n \cup \{(n, G(n))\}$ is a set.

Now, we can define the function G as follows:

$$G(n) = y \iff (n = 0 \land y = A_0)$$
$$\lor [(\exists g)[Dmn(g) = n + 1 \land g(0) = A_0 \land g(n) = y$$
$$\land (\forall i < n)g(i + 1) = F(i, g(i))].$$

This definition succeeds since, for each n, the desired function g is just $G \restriction n$, which we have shown to be a set. Finally, the image $G[\omega]$ is a set by the Axiom of Replacement. □

Example 4.3.14. Let $G(0) = 0$ and let $G(n + 1) = n + 1 + G(n)$. Then $G(n) = 1 + 2 + \cdots + n = \sum_{i \leq n} i$. It can be shown by induction that this equals $\frac{n(n+1)}{2}$.

Here is a key example for set theory.

Example 4.3.15. The finite part of the hierarchy of sets may be defined using the power set as follows. Define the function V by $V(0) = 0 = \emptyset$ and, for each n, $V(n + 1) = \mathcal{P}(V(n))$. We usually write V_n for $V(n)$. Thus, we have $V_1 = \{0\}$, $V_2 = \{0, \{0\}\} = \{0, 1\}$, $V_3 = \{0, 1, \{1\}, \{0, 1\}\}$, and so on. For the resulting class function V, $\bigcup Rng(V) = \bigcup_n V(n)$, which we denote as V_ω. We see later that this is the same class as the family HF of hereditarily finite sets discussed in Chapter 3. We are tempted to say that V_ω is the smallest set which is closed under the Power Set operator. The problem is that no axiom guarantees the existence of any set which is closed under Powerset. It is the class of all sets, not a set itself, which is the natural family closed under powerset. Now, in order to define ω as the least set containing 0 and closed under successor, we had to have the Axiom of Infinity to say that there some such set. So, do we need *another* axiom which says that some set exists which is closed under powerset? This seems inelegant. Fortunately, the answer is *NO*. The existence of ω, together with the other axioms of ZF, will produce a bountiful family of transfinite sets. Here we can just use Theorem 4.3.13 to obtain the set V_ω. That is, let $A_0 = \emptyset$ and let $F(n,x) = \mathcal{P}(x)$. Then applying Theorem 4.3.13, we have $G(0) = \emptyset = V_0$ and, for each n, $G(n+1) = F(n, V_n) = \mathcal{P}(V_n) = V_{n+1}$. So, G is a class function with $G(n) = V_n$ and thus $G[\omega] = \{V_n : n \in \omega\}$ is a set and hence $\bigcup G[\omega] = V_\omega$ is also a set, as desired. In Chapter 5, the hierarchy V_α of sets is extended to the transfinite.

We may now revisit the notion of a monotone class operator. This result extends Proposition 4.3.5 and Theorem 4.3.8 from set operators to class operators.

Theorem 4.3.16.

1. *For any monotone class operator* Γ, *$\{(n,x) : x \in \Gamma^n\}$ is a set and Γ^ω is a set.*
2. *If* Γ *is a finitary monotone class operator, then Γ^ω is the least fixed point of Γ.*

Proof. To prove (1), let $G(0) = \emptyset$ and let $G(n+1) = \Gamma(G(n))$ for each n. Then G is a class function with domain ω by Theorem 4.3.13 and $G(n) = \Gamma^n$. Moreover, $\Gamma^\omega = \bigcup Rng(G)$. For (2), the proof that Γ^ω is the least fixed point is the same as in Theorem 4.3.8. □

It is frequently useful to have other variables as part of a recursive definition. Some examples are given as follows:

Theorem 4.3.17. *Suppose that F and H are class functions such that $H(x)$ is defined for all sets x and such that $F(n,x,y)$ is defined for all $n \in \omega$ and for all sets x and y. Then there is a unique class function G such that $G(n,x)$ is defined for all $n \in \omega$ and all sets x, such that, for every $n \in \omega$ and every set x, $G(0,x) = H(x)$, and $G(n+1,x) = F(n,x,G(n,x))$.*

Proof. The proof is a small modification of the argument in Theorem 4.3.13 which treats x as a parameter. We show that, for each n, $G(n,x)$ is a set and that the restriction $G_x \upharpoonright n = \{(i, G(i,x)) : i, n\}$ is a set. Then we have the following definition of G:

$$G(n,x) = y \iff (n = 0 \,\wedge\, y = H(x))$$
$$\vee\; [(\exists g)[Dmn(g) = n+1 \,\wedge\, g(0) = H(x) \,\wedge\, g(n) = y$$
$$\wedge\; (\forall i < n)g(i+1) = F(i,x,g(i))].$$

Details are left to the exercises. □

Example 4.3.18. Let $G(0, x) = 0$ and let $G(n+1, x) = x + G(n, x)$ for all $n, x \in \omega$. Then $G(n, x) = n \cdot x$. Now, we can define the factorial function by letting $F(0) = 1$ and $F(n+1) = (n+1) \cdot F(n)$.

Example 4.3.19. Let $G(0, x) = 1$ and let $G(n+1, x) = x \cdot G(n, x)$. Then $G(n, x) = x^n$.

Another key example in set theory is the definition of the transitive closure. We first need to define the notion of a transitive set.

Definition 4.3.20. A set x is *transitive* if for every $y \in x$ and every $z \in y$, $z \in x$ holds.

For example, ω is transitive, and every natural number is transitive by Theorem 4.1.5. An example of a non-transitive set is $\{\{0\}\}$. We show in Chapter 5 that every set belongs to a transitive set.

Definition 4.3.21. Let S be a set. The *transitive closure* of a, $\mathtt{trcl}(a)$, is the inclusion-smallest transitive set containing a as an element. That is, $\mathtt{trcl}(a)$ is the least fixed point of the finitary monotone operator Γ_a, where $\Gamma_a(X) = \{a\} \cup \bigcup X$.

Note here that there is at this point no universal transitive set U such that $a \in U$, which would enable us to ensure that this least fixed point exists. For this, we use the general result about finitary monotone operators.

Lemma 4.3.22. Γ_a *is a finitary monotone operator for any set a.*

Proof. This is left to the exercises. $\qquad\square$

Theorem 4.3.23. *For any set a, $\mathtt{trcl}(a)$ is a set.*

Proof. This is immediate from Theorem 4.3.16 and Lemma 4.3.22. $\qquad\square$

Finally, we can show that $\mathtt{trcl}$ is a class function.

Theorem 4.3.24. *There is a class function F such that $F(a) = \mathtt{trcl}(a)$ for all sets a.*

Proof. Define the class function G such that $G(0, a) = a$ for all sets a and $G(n+1, a) = \Gamma_a(G(n, a)) = \{a\} \cup \bigcup G(n, a)$. Then G is a class function by Theorem 4.3.17, and it is easy to see that $G(n, a) = \Gamma_a^n$ for all n. Then by Lemma 4.3.22, $F(a) = \mathtt{trcl}(a) = \Gamma_a^\omega = \bigcup \{G(n, a) : n \in \omega\}$. $\qquad\square$

Corollary 4.3.25 (Axiom of Regularity for classes). *Let C be a non-empty class. There is an element $x \in C$ such that no element of x belongs to C.*

Proof. Let y be any element of C. Consider the non-empty set $C \cap \text{trcl}(y)$. The fact that this is indeed a set and not just a class follows from Exercise 3.5.1. Use the Axiom of Regularity to find an $\in$-minimal element x of $C \cap \text{trcl}(y)$. All elements of x belong to $\text{trcl}(y)$, and so by the minimal choice of x, none of them can belong to C. Thus, the set x works as required. $\qquad\square$

Let us return to the hierarchy V_n of sets defined in Example 4.3.15.

Proposition 4.3.26.

1. *For each $n \in \omega$, $V_n \subseteq V_{n+1}$.*
2. *For each $n \in \omega$, V_n is transitive.*
3. *V_ω is transitive.*

Proof. The proof of part 1 is by induction on n.

Base Step. $n = 0$. Then $V_0 = \emptyset \subseteq V_1$.

Successor Step. Suppose by induction that $V_n \subseteq V_{n+1}$. Let $x \in V_{n+1}$. Then by definition of V_{n+1}, $x \subseteq V_n$. Thus, by the induction hypothesis, $x \subseteq V_{n+1}$, and therefore $x \in V_{n+1}$. This shows that $V_{n+1} \subseteq V_{n+2}$.

For the proof of part 2, suppose that $x \in y \in V_n$. Then by definition, $y \subseteq V_{n-1}$ and hence $x \in V_{n-1}$. It now follows from part 1 that $x \in V_n$.

For the proof of part 3, suppose that $x \in y \in V_\omega$, then $x \in y \in V_n$ for some n. Thus, by part 2, $x \in V_n$ and therefore $x \in V_\omega$. $\qquad\square$

Finally, we return to course-of-values recursive definitions. These are very important for transfinite recursive definitions in the coming Chapter 5.

Theorem 4.3.27 (Course-of-Values Recursive Definitions).

1. *Suppose that F is a class function such that $F(x)$ is defined for every set x. Then there is a unique class function G such that $\text{Dmn}(G) = \omega$ and for every $n \in \omega$, $G(n) = F(G \restriction n)$. Therefore, the image $G[\omega]$ is a set.*

2. *Suppose that F is a class function such that $F(x, y)$ is defined for all sets x and y. Then there is a unique class function G such that for all $n \in \omega$ and all sets x, $G(n, x) = F(x, G_x \upharpoonright n)$, where $G_x \upharpoonright n = \{(i, G(i, x)) : i < n\}$.*

Proof. Here is the proof of part (1). Let the class function F be given so that $F(x)$ is defined for every set x and consider the proposed function G such that $G(n) = F(G \upharpoonright n)$. Now, consider the recursive definition of the function $\hat{G}$ such that $\hat{G}(n) = G \upharpoonright n$. We see that

1. $\hat{G}(0) = \emptyset$;
2. $\hat{G}(n+1) = \hat{G}(n) \cup \{(n, G(n)\} = \hat{G}(n) \cup \{(n, F(\hat{G}(n)))\}$.

It now follows from Theorem 4.3.13 that $\hat{G}$ is a class function with domain ω and $G(n) = \hat{G}(n+1)(n)$ for all $n \in \omega$.

For part (2), consider the function $\hat{G}$ such that $\hat{G}(n, x) = G_x \upharpoonright n$ for $n \in \omega$ and proceed as in part (1). $\qquad \square$

Exercises for Section 4.3

Exercise 4.3.1. Show that if A is transitive for every set $A \in S$, then $\bigcup S$ is transitive.

Exercise 4.3.2. Show that the union operator is monotone and finitary.

Exercise 4.3.3. Show that for any monotone operator Γ on a set A and any $m < n$, $\Gamma^m \subseteq \Gamma^n$.

Exercise 4.3.4. Show that a set x is transitive if and only if $\bigcup x \subseteq x$.

Exercise 4.3.5. Show that for any sets A and B, if $A \in B$ or if $A \subseteq B$, then $\mathtt{trcl}(A) \subseteq \mathtt{trcl}(B)$.

Exercise 4.3.6. Prove that if A is a transitive set, then $\mathcal{P}(A)$ is transitive.

Exercise 4.3.7. Show that if every element of a set A is transitive, then $\bigcup A$ is transitive.

Exercise 4.3.8. Show that $\{A\} \cup \bigcup_{a \in A} \mathtt{trcl}(a)$ is transitive for any set A.

Exercise 4.3.9. Use inductive definability to show that $\{x \in \mathbb{N} : x = 3 \ (mod\ 7)\}$ is a set.

Exercise 4.3.10. Prove that the powerset operator is monotone, that is, $A \subseteq B \Longrightarrow \mathcal{P}(A) \subseteq \mathcal{P}(B)$.

Exercise 4.3.11. Prove that for all $m, n \in \mathbb{N}$, $m + n = n + m$.
Hint: Use double induction and let $\phi(m)$ be $(\forall n)\ m + n = n + m$. Then for each $m > 0$, prove $\phi(m)$ by induction on n.

Exercise 4.3.12. Prove by induction on z that for all $x, y, z \in \mathbb{N}$: $(x + y) + z = x + (y + z)$.

Exercise 4.3.13. Use inductive definability to show that $\{(x, y, z) \in \mathbb{N} : x \cdot y = z\}$ is a set. *Hint*: $x \cdot 0 = 0$ and $x \cdot Sy = x \cdot y + x$ for all x, y.

Exercise 4.3.14. Prove that for all n, $1 \cdot n = n$.

Exercise 4.3.15. Prove by induction on z that for all $x, y, z \in \mathbb{N}$, $(x + y) \cdot z = x \cdot z + y \cdot z$.

Exercise 4.3.16. Let $\Gamma(X) = X \cup \{3, 7\} \cup \{x + y : x, y \in X\}$ for $X \in \mathcal{P}(\mathbb{N})$. Compute $\Gamma^1, \Gamma^2, \Gamma^3$ and determine the least fixed point M of Γ. Prove that $2 \notin M$ and that every number $n \geq 15$ is in M. *Hint*: Consider n *modulo* 3.

Exercise 4.3.17. Fill in the details in the proof of Theorem 4.3.17.

Exercise 4.3.18. Show that the class V_ω is equal to the smallest set containing 0 and closed under pairing and pairwise union.

4.4 Cardinality

In this section, we develop the basic features of the set-theoretic notion of size, referred to as cardinality.

Definition 4.4.1. Let x, y be sets. Say that x, y have the same *cardinality*, in symbols $|x| = |y|$, if there is a bijection $f : x \to y$. Say that $|x| \leq |y|$ if there is an injection from x to y.

Theorem 4.4.2. *Having the same cardinality is an equivalence relation and $\leq$ is a quasiorder, that is, reflexive and transitive.*

The proofs are left as exercises.

Theorem 4.4.3 (Schröder–Bernstein). *If $|A| \le |B|$ and $|B| \le |A|$, then A, B have the same cardinality.*

Proof. Let A, B be sets and $f : A \to B$ and $g : B \to A$ be injections. We construct a bijection $h : A \to B$. We observe that the functions f and g create *orbits* for the elements of $A \cup B$. An orbit is a sequence $(\ldots, a_i, b_i, a_{i+1}, b_{i+1}, \ldots)$ such that, for each i, $b_i = f(a_i)$ and $a_{i+1} = g(b_i)$. There are three possible types of orbits. For any $a \notin g[B]$, there is an orbit $(a_0, b_0, a_1, \ldots)$ of type ω beginning with $a_0 = a$. Similarly, for any $b \notin f[A]$, there is an orbit of type ω beginning with $b_0 = b$. Finally, there are orbits $(\ldots, a_{-1}, b_{-1}, a_0, b_0, a_1, \ldots)$ of type $\mathbb{Z}$ where each $a_i \in g[B]$ and each $b_i \in f[A]$. The orbits partition $A \cup B$, so we may define h on each orbit, as follows. For the orbits $(b_0, a_1, \ldots)$ beginning with b_0, we define $h(a_{i+1}) = b_i = g^{-1}(a_{i+1})$. For the other two types of orbits, we define $h(a_i) = b_i = f(a_i)$. It is easy to check that this is a bijection from A onto B. $\qquad\square$

Theorem 4.4.4. *Distinct natural numbers have distinct cardinalities.*

Proof. It will be enough to show that if x, y are finite sets and $y \subseteq x$ and $y \ne x$, then y, x have distinct cardinalities. Suppose for contradiction that this fails for some x, y. Let $a = \{z \subseteq x : |z| = |x|\}$. The set $a \subseteq \mathcal{P}(x)$ is certainly non-empty, containing at the very least the set x itself. Let $z \in a$ be a $\subseteq$-minimal element. Note that $z \ne x$ since $y \in a$ and y is a proper subset of x. Let $h : x \to z$ be a bijection, and let $u = h[z]$, the image of z under h. Then $u \subseteq z$ and $|u| = |z|$, since $h \restriction z : z \to u$ is a bijection. Moreover, $u \ne z$: if i is any element of the non-empty set $x \setminus z$, then $h(i)$ belongs to $z \setminus u$. Thus, u is a proper subset of z which has the same cardinality of z and so the same cardinality as x. This contradicts the $\subseteq$-minimal choice of the set z. $\qquad\square$

This theorem completely determines the possible cardinalities of finite sets. Every finite set has the same cardinality as some natural number by Theorem 4.2.8, and distinct natural numbers have distinct cardinalities. Thus, the cardinalities of finite sets are linearly ordered. One can ask if this feature persists even for infinite cardinalities. The answer depends on the Axiom of Choice. Assuming the Axiom of Choice, we show that even the infinite cardinalities are linearly ordered.

If $|A| = |n|$ for some $n \in \omega$, then we write $|A| = n$ and say that A has cardinality n.

Let $A \oplus B = \{(0, a) : a \in A\} \cup \{(1, b) : b \in B\}$; this is the *disjoint union* of A and B.

Proposition 4.4.5. *Suppose that A and B are two sets such that $|A| = m$ and $|B| = n$ for some $m, n \in \omega$. Then*

1. $|A \oplus B| = m + n$;
2. $|A \times B| = m \cdot n$;
3. $|2^A| = 2^m$.

Proof. (1) Let $f : A \to m = \{0, 1, \ldots, m - 1\}$ and $g : B \to n$ be bijections and define $h : A \oplus B \to m + n$ by $h(0, x) = x$ and $h(1, x) = m + x$.

(2) is left as an exercise.

(3) Since $|A| = m$, it suffices to show that $|2^{\{0,1,\ldots,m-1\}}| = 2^m$. Given $x \in 2^m$, recall that x is a function from $\{0, 1 \ldots, m - 1\}$ to $\{0, 1\}$. Let $f(x) = \sum_{i < m} x(i) 2^{m-i}$. This will be a bijection from $2^{\{0,1,\ldots,m-1\}}$ to $2^m = \{0, 1, \ldots, 2^m - 1\}$. $\square$

If $(A, \leq_A)$ and $(B, \leq_B)$ are partial orderings (possibly linear or well-orderings), then there are natural orderings on $A \oplus B$, $A \times B$, and 2^A. For $A \oplus B$, let $(0, a_1) \leq (0, a_2) \iff a_1 \leq_A a_2$, let $(1, b_1) \leq (1, b_2) \iff b_1 \leq_B b_2$, and let $(0, a) < (1, b)$ for all $a \in A$ and $b \in B$. If A is isomorphic to $(m, <)$ and B is isomorphic to $(n, <)$, then $A \oplus B$ will be order isomorphic to $(m + n, <)$, using the map given above in Proposition 4.4.5.

For $A \times B$, we use the lexicographic ordering so that $(a_1, b_1) \leq (a_2, b_2) \iff a_1 <_A a_2 \vee (a_1 = a_2 \wedge b_1 \leq_B b_2)$. Then the natural mapping of $A \times B$ to $m \cdot n$ will be an order isomorphism.

For 2^A, we again use the lexicographic order so that $x < y$ if and only if $x(i) < y(i)$, where i is the least such that $x(i) \neq y(i)$.

Even when A and/or B are infinite sets, if A and B are linearly ordered, or well ordered, then $A \oplus B$ will be linearly (well ordered), and similarly for $A \times B$. However, if A is infinite, then 2^A will not be well ordered. (See the exercises.)

We can generalize some of the rules of addition, multiplication, and exponentiation, as follows:

Proposition 4.4.6. *For any sets A, B, and C,*

1. $|A \oplus B| = |B \oplus A|$ *and* $|A \times B| = |B \times A|$;
2. $|A \oplus (B \oplus C)| = |(A \oplus B) \oplus C|$ *and* $|A \times (B \times C)| = |(A \times B) \times C|$;
3. $|A \times (B \oplus C)| = |(A \times B) \oplus (A \times C)|$.

Proof. For the multiplicative parts of (1) and (2), we have natural isomorphisms mapping (a, b) to (b, a) and mapping $((a, (b, c))$ to $((a, b), c)$. The other parts are left as exercises. □

Proposition 4.4.7. *For any sets A, B, and C,*

1. $|A^{B \oplus C}| = |A^B \times A^C|$;
2. $|A^C \times B^C| = |(A \times B)^C|$;
3. $|(A^B)^C| = |A^{B \times C}|$.

Proof. (1) Let $(F, G) \in A^B \times A^C$. Then $F : B \to A$ and $G : C \to A$. Then we can map (F, G) to the function $H : B \oplus C \to A$ defined by $H(0, b) = F(b)$ and $H(1, c) = G(c)$. To check that this is one-to-one, suppose that (F_1, G_1) and (F_2, G_2) map to the same function H. Then for any $b \in B$, $F_1(b) = H(0, b) = F_2(b)$ and for any $c \in C$, $G_1(c) = H(1, c) = G_2(c)$, so that $F_1 = F_2$ and $G_1 = G_2$. To check that this is surjective, let $H : B \oplus C \to A$ be given. Then we may define $F : B \to A$ by $F(b) = a \iff H(0, b) = a$ and $G : C \to A$ by $G(c) = a \iff H(1, c) = c$.
Parts (2) and (3) are left as exercises. □

We conclude this section by proving that there are many distinct infinite cardinalities.

Theorem 4.4.8 (Cantor). *For every set x,* $|x| \leq |\mathcal{P}(x)|$ *and* $|x| \neq |\mathcal{P}(x)|$.

Proof. Clearly, $|x| \leq |\mathcal{P}(x)|$ since the function $f : y \mapsto \{y\}$ is an injection from x to $\mathcal{P}(x)$.

To show that $|x| \neq |\mathcal{P}(x)|$ suppose for contradiction that x is a set and $f : x \to \mathcal{P}(x)$ is any function. It will be enough to show that $Rng(f) \neq \mathcal{P}(x)$, ruling out the possibility that f is a bijection. Consider the set $y = \{z \in x : z \notin f(z)\}$; we show that $y \notin Rng(f)$. For contradiction, assume that $y \in Rng(f)$ and fix $z \in x$ such that

$y = f(z)$. Consider the question whether $z \in y$. If $z \in y$, then $z \notin f(z)$ by the definition of y, and then $z \notin y = f(z)$. If, on the other hand, $z \notin y$, then $z \in f(z)$ by the definition of y, and so $z \in y = f(z)$. In both cases, we have arrived at a contradiction. $\qquad\square$

Thus, $\mathcal{P}(\omega)$ has strictly greater cardinality than ω, $\mathcal{P}\mathcal{P}(\omega)$ has strictly greater cardinality than $\mathcal{P}(\omega)$, and so on. We have produced infinitely many infinite sets with pairwise distinct cardinalities.

Exercises for Section 4.4

Exercise 4.4.1. Suppose that A and B are two sets such that $|A| = m$ and $|B| = n$ for some $m, n \in \omega$. Show that $|A \times B| = m \cdot n$.

Exercise 4.4.2. Suppose that A and B are two sets such that $|A| = m$ and $|B| = n$ for some $m, n \in \omega$. Show that $|B^A| = n^m$. *Hint:* Use the representation of a natural number in base n.

Exercise 4.4.3. Prove that for any sets A and B, if A and B are linearly ordered (well ordered), then $A \oplus B$ will be linearly (well ordered) under the ordering given above.

Exercise 4.4.4. Prove that for any sets A and B, if A and B are linearly ordered (well ordered), then $A \times B$ will be linearly (well ordered) under the ordering given above.

Exercise 4.4.5. Show that for any sets A and B, $|A \oplus B| = |B \oplus A|$.

Exercise 4.4.6. Show that for any sets A, B, and C, $|A \oplus (B \oplus C)| = |(A \oplus B) \oplus C|$.

Exercise 4.4.7. Show that for any sets A, B, and C, $|A \times (B \oplus C)| = |(A \times B) \oplus (A \times C)|$.

Exercise 4.4.8. Show that for any sets A, B, and C, $|A^C \times B^C| = |(A \times B)^C|$.

Exercise 4.4.9. Show that for any sets A, B, and C, $|(A^B)^C)| = |(A^{(B \times C)}|$.

Exercise 4.4.10. Show that for disjoint sets B, C, $|A^B \times A^C| = |A^{(B \cup C)}|$.

Exercise 4.4.11. Show that the lexicographic order on 2^ω is a linear ordering but is not well founded.

Exercise 4.4.12. Prove that if $|x| \le |y|$, then $|\mathcal{P}(x)| \le |\mathcal{P}(y)|$.

Exercise 4.4.13. Prove that for every set x, $|\mathcal{P}(x)| = |2^x|$.

Exercise 4.4.14. Prove that having the same cardinality is an equivalence relation.

Exercise 4.4.15. Show that the relation "$|x| \le |y|$" is a quasiorder, that is, it is transitive and reflexive.

Exercise 4.4.16. Use the Axiom of Choice to prove that if y is a surjective image of x then $|y| \le |x|$.

Exercise 4.4.17. Prove that whenever x is a set, then there is a set y such that $|z| < |y|$ for every $z \in x$.

4.5 Countable and Uncountable Sets

The most important cardinality-related concept in mathematics is countability. We use it in this section to provide the scandalously easy proof of the existence of transcendental real numbers discovered by Cantor.

Definition 4.5.1. A set x is *countable* if $|x| \le |\omega|$. A set which is not countable is *uncountable*.

As a matter of terminology, some authors require countable sets to be infinite. By the following theorem, this restricts the definition to the collection of sets which have the same cardinality as ω.

Theorem 4.5.2.

1. *If x is countable, then either x is finite or $|x| = |\omega|$.*
2. *A non-empty set is countable if and only if it is a surjective image of ω.*
3. *A surjective image of a countable set is countable.*
4. *(With the Axiom of Choice) The union of a countable collection of countable sets is again countable.*
5. *The product of two countable sets is countable.*

Proof. For (1), we first argue that for every set $x \subseteq \omega$, either x is finite or $|x| = |\omega|$. This is easy to see though: If the set $x \subseteq \omega$ is infinite, then its enumeration in increasing order is a bijection between ω and x.

Now, suppose that x is an arbitrary countable set, and choose an injection $f : x \to \omega$. Let $y = Rng(f)$, so $f : x \to y$ is a bijection. From the first paragraph, the set y is either finite or has the same cardinality as ω, and so the same has to be true about x. This completes the proof of (1).

For (2), if $f : \omega \to x$ is a surjection of ω onto any set x, then the function $g : x \to \omega$ defined by $g(z) = \min\{n \in \omega : f(n) = z\}$ is an injection of x to ω, confirming that x is countable. On the other hand, if x is countable, then either x is infinite and then x is in fact a bijective image of ω by (1), or x is finite and then it is a bijective image of some natural number n. Any extension of this bijection to a function from ω to x will be a surjection.

For (3), let x be a countable non-empty set and $f : x \to y$ be a surjection. By (2), there is a surjection $g : \omega \to x$ and then $f \circ g$ will be a surjection of ω onto y, confirming the countability of x.

For (4), we first show (without AC) a special case: The set $\omega \times \omega$ is countable. Indeed, one bijection between $\omega \times \omega$ and ω is the Cantor pairing function defined by $f(m, n) = \frac{1}{2}(n + m)(n + m + 1) + m$; we denote $f(m, n)$ as $[m, n]$. Now, suppose that $b = \{a_i : i \in \omega\}$ is a countable set, all of whose elements are again countable. To show that $\bigcup b$ is countable, we produce a surjection from $\omega \times \omega$ to $\bigcup b$, which in view of the special case and (3) show that $\bigcup b$ is countable.

For every $i \in \omega$, let c_i be the set of all surjections from ω to a_i. Since each a_i is assumed to be countable, each set c_i is non-empty. Use the Axiom of Choice to find a selector h: a map with domain ω such that for every $i \in \omega$, $h(i) \in c_i$. Let $f : \omega \times \omega \to \bigcup b$ be the map defined by $f(i, j) = h(i)(j)$. This is the desired surjection.

The proof of part (5) is left as an exercise. $\qquad \square$

Item (4) in its generality cannot be proved without the Axiom of Choice. Lebesgue, an opponent of AC, used item (4) unwittingly to develop his theory of integration; any such a theory has to use some form of (4).

Theorem 4.5.3. *The following sets are countable:*

1. *the set of integers;*
2. *the set $A^{<\omega}$ of all finite sequences of elements of any countable set A;*
3. *the set of rational numbers;*
4. *the set of all open intervals with rational endpoints;*
5. *the set of all polynomials with integer coefficients;*
6. *the set of all algebraic numbers.*

Proof. We discuss part (2) and leave the others as exercises. First, we see by induction that A^n is countable for each n. That is, $A^1 = A$ is countable. Assuming that A^n is countable, then $A^{n+1} = A \times A^n$ is countable as a product of countable sets. Finally, $A^{<\omega} = \bigcup_n A^n$ is countable as a countable union of countable sets. □

Theorem 4.5.4. $|\mathbb{R}| = |\mathcal{P}(\omega)|$.

While we have not developed the real numbers $\mathbb{R}$ formally, any usual concept of real numbers will be sufficient to prove this theorem.

Proof. By the Schröder–Bernstein theorem, it is enough to provide an injection from $\mathbb{R}$ to $\mathcal{P}(\omega)$ as well as an injection from $\mathcal{P}(\omega)$ to $\mathbb{R}$.

To construct an injection from $\mathbb{R}$ to $\mathcal{P}(\omega)$, we construct an injection f from $\mathbb{R}$ to $\mathcal{P}(\mathbb{Q})$ and finish the argument by Theorem 4.5.2(1). For any real r, let $f(r) = \{q \in \mathbb{Q} : q < r\}$; this set is part of the *Dedekind cut* corresponding to r. To see that this is an injection, let $r < s$ be two distinct real numbers. Then, by the density of the rationals in $\mathbb{R}$, there is a rational q such that $r < q < s$. Thus, $q \in f(s)$, but $q \notin f(r)$ and hence $f(r) \neq f(s)$.

To construct an injection from $\mathcal{P}(\omega)$ to $\mathbb{R}$, consider the function $g : \mathcal{P}(\omega) \to \mathbb{R}$ defined by the following formula: $g(y)$ is the unique element of the closed interval $[0, 1]$ whose ternary expansion consists of 0's and 2's only, and $n+1$-st digit of the ternary expansion of $g(y)$ is 2 if $n \in y$, and the $n + 1$-st digit is 0 if $n \notin y$. It is easy to check that this is an injection.The image of g is often referred to as the Cantor set. □

Corollary 4.5.5 (Cantor). *There is a real number which is not the root of a non-zero polynomial with integer coefficients.*

Proof.   The set $\mathcal{P}(\omega)$ is uncountable by Theorem 4.4.8, and so is $\mathbb{R}$. On the other hand, the set of algebraic real numbers is countable by Theorem 4.5.3(6). Thus, there must be a real number which is not algebraic. $\square$

The presented proof is incomparably easier than any proof that a specific real number (say π or e) is not algebraic. Also, it does not use almost any knowledge about real numbers.

Exercises for Section 4.5

Exercise 4.5.1. Let x be a countable set. Show that any set consisting of pairwise disjoint subsets of x is countable.

Exercise 4.5.2. Show, without appealing to the Axiom of Choice, that for any non-empty sets A and B, if there is an injection from A into B, then there is a surjection from B onto A.

Exercise 4.5.3. Show that if A is countable, then any subset of A is countable.

Exercise 4.5.4. Show that the product $A \times B$ of two countable sets is countable. *Hint*: Use the pairing function $[m, n] = \frac{1}{2}(n + m)(n + m + 1) + m$.

Exercise 4.5.5. Prove directly that $\mathbb{N}^{<\omega}$ is countable, without appealing to the Axiom of Choice.

Exercise 4.5.6. Show that if A is countable, then for any b, $A \cup \{b\}$ is countable.

Exercise 4.5.7. Show that the family of finite subsets of ω is countable.

Exercise 4.5.8. Show that the set $\mathcal{C}(\mathbb{R})$ of continuous real functions has the same cardinality as $\mathbb{R}$. *Hint*: A continuous function is determined by its values on the rational numbers.

Exercise 4.5.9. Show that $|2^\omega \times 2^\omega| = |2^\omega|$ and therefore $|\mathbb{R} \times \mathbb{R}| = |\mathbb{R}|$.

Exercise 4.5.10. Show that $|\mathbb{R}^\omega| = |\mathbb{R}|$.

Chapter 5

Ordinal Numbers and the Transfinite

In this chapter, we show that the processes of induction and recursion can be extended far beyond ω. It is exactly the use of this extended, transfinite induction, and recursion that sets set theory apart from other field of mathematics. While the idea may sound far-fetched at first, it is very powerful and has found many uses in mathematics: the equivalence of Zorn's lemma and the Axiom of Choice, the Cantor–Bendixson analysis of closed sets of reals, or the stratification of Borel sets of reals into a hierarchy can serve as good examples. We demonstrate the existence of uncountable ordinals, using the Axiom of Replacement.

5.1 Ordinals

We first define the von Neumann *ordinal numbers*. This is a natural extension of the notion of von Neumann natural numbers. That is, the set ω of natural numbers is the first transfinite ordinal, followed by $\omega + 1 = \omega \cup \{\omega\}$, $\omega + 2$, and so on. Each new larger ordinal is simply the set of all of the smaller ordinals. Ordinals are typically denoted by lower case Greek letters such as $\alpha, \beta, \gamma, \ldots$ The collection of ordinals is itself naturally linearly ordered: Given two ordinals α, β, then either α is an initial segment of β or vice versa, β is an initial segment of α. Note that while ω and $\mathbb{N}$ both refer to the set of natural numbers, we use the notation ω when we are discussing ordinals.

Definition 5.1.1. A set x is an *ordinal number*, or ordinal for short, if it is transitive and linearly ordered by $\in$.

In particular, every natural number as well as ω is an ordinal. There are other ordinals as well, e.g. $\omega \cup \{\omega\}$. If we want to develop the theory of ordinals without the Axiom of Regularity, the definition needs to be amended: A set x is an ordinal if it is transitive, linearly ordered by $\in$, and every subset of x has an $\in$-minimal element (the last clause is automatic if the Axiom of Regularity is present). Every ordinal with the membership relation is a linear ordering as per the definition, and we always view ordinals as linear orderings.

We first record the most useful technical properties of ordinals.

Theorem 5.1.2. *Let α be an ordinal:*

1. *Every element of α is an ordinal.*
2. *Every $\in$-initial segment is either an element of α or equal to α.*

Proof. For (1), let $\beta \in \alpha$. We have to verify that β is linearly ordered by $\in$ and transitive. To show linearity, observe that $\beta \subseteq \alpha$ by the transitivity of α, and as α is linearly ordered by $\in$, so is β. To show transitivity, suppose that $\gamma \in \beta$ and $\delta \in \gamma$; we must conclude that $\delta \in \beta$. By the transitivity of α, all β, γ, δ are in α. Since $\in$ is a linear ordering on α and $\delta \in \gamma \in \beta$, $\delta \in \beta$ follows as required.

For (2), let $x \subseteq \alpha$ be an $\in$-initial segment of α and $x \neq \alpha$; we must argue that $x \in \alpha$. Let $\beta \in \alpha$ be the $\in$-minimal element of the non-empty set $\alpha \setminus x$ obtained by the Axiom of Regularity. We show that $\beta = x$ and that completes the proof as $\beta \in \alpha$.

For the inclusion $\beta \subseteq x$, choose $\gamma \in \beta$ and argue that γ must be an element of x. If this failed, γ would be an element of $\alpha \setminus x$ $\in$-smaller than β, contradicting the minimal choice of β. For the inclusion $x \subseteq \beta$, choose $\gamma \in x$ and argue that γ must be an element of β. If this failed, then by the linearity of $\in$ on the ordinal α, it would have to be the case that $\beta = \gamma$ or $\beta \in \gamma$. The former case is impossible as $\beta \notin x$ and $\gamma \in x$. In the latter case, note that x is an $\in$-initial segment of α and so $\beta \in \gamma$ and $\gamma \in x$ implies $\beta \in x$, again contradicting the assumption that $\beta \notin x$. The proof is complete. $\square$

Now, it is time to prove the more salient features of ordinal numbers.

Theorem 5.1.3 (Linear Ordering). *The class of ordinals is transitive and is linearly ordered by $\in$.*

Proof. It is easy to see that $\in$ is a transitive relation on ordinals. If $\gamma \in \beta \in \alpha$ are ordinals, then, as α is a transitive set, $\gamma \in \alpha$ must hold, and the transitivity of $\in$ has been proved. Let Ord be the class of ordinals. It is immediate from Theorem 5.1.2 that Ord is transitive. That is, if $\alpha \in \beta \in Ord$, then $\alpha \in Ord$ as well.

For the linearity, let α, β be ordinals; we have to argue that either $\alpha \in \beta$ or $\beta \in \alpha$ or $\alpha = \beta$ holds. To this end, consider the set $\gamma = \alpha \cap \beta$. As an intersection of two transitive sets, it is transitive and therefore an $\in$-initial segment of both α and β.

Note that it must be the case that either $\gamma = \alpha$ or $\gamma = \beta$ holds. If both of these equalities failed, then both $\gamma \in \alpha$ and $\gamma \in \beta$ must hold by Theorem 5.1.2. But then $\gamma \in \alpha \cap \beta$, so $\gamma \in \gamma$ by the definition of γ, and this contradicts the Axiom of Regularity.

Thus, γ is equal to one of the ordinals α, β; say that $\gamma = \alpha$. If $\gamma = \beta$ then we conclude that $\alpha = \beta$. If $\gamma \neq \beta$, then by Theorem 5.1.2(2), $\gamma \in \beta$ and we conclude that $\alpha \in \beta$. In both cases, the linearity of $\in$ is confirmed. $\square$

Corollary 5.1.4. *The class of ordinals is not a set.*

Proof. Assume for contradiction that the class of ordinals is a set x. It follows from Theorem 5.1.3 that the set x is transitive and linearly ordered by $\in$. Therefore, x is an ordinal, and so $x \in x$, contradicting the Axiom of Regularity. $\square$

Theorem 5.1.5 (Rigidity). *Whenever α, β are ordinals and $i : \alpha \to \beta$ is an isomorphism of linear orders, then $\alpha = \beta$ and $i = \mathrm{id}$.*

Proof. Assume that α, β are ordinals and $i : \alpha \to \beta$ is an isomorphism. Suppose for contradiction that i is not the identity. Then, there must be an ordinal $\gamma \in \alpha$ such that $i(\gamma) \neq \gamma$. Use the Axiom of Regularity to choose the $\in$-least ordinal $\gamma \in \alpha$ such that $i(\gamma) \neq \gamma$. Since $i(\gamma) \in \beta$, $i(\gamma)$ is an ordinal by Theorem 5.1.2(1). By linearity (Theorem 5.1.3), there are three possible cases: either $i(\gamma) = \gamma$, or $i(\gamma) \in \gamma$, or $\gamma \in i(\gamma)$. We reach a contradiction in each case.

First, $i(\gamma) = \gamma$ is impossible as γ was chosen precisely so that $i(\gamma) \neq \gamma$. Second, assume that $i(\gamma) \in \gamma$. By the minimal choice of γ, the equality $i(i(\gamma)) = i(\gamma)$ must hold. This means that the distinct ordinals γ and $i(\gamma)$ are sent by the isomorphism i to the same value $i(\gamma)$, which is a contradiction. Third, assume that $\gamma \in i(\gamma)$. In this case, since β is a transitive set and it contains $i(\gamma)$, it must contain also its element γ. Let $\delta \in \alpha$ be an element such that $i(\delta) = \gamma$. Since i is an isomorphism of linear orders and $i(\delta) = \gamma \in i(\gamma)$, it must be the case that $\delta \in \gamma$. By the minimality choice of γ, $i(\delta) = \delta \neq \gamma$, a final contradiction. $\qquad\square$

Here is an alternative definition of an ordinal.

Definition 5.1.6. The set x is said to be *hereditarily transitive* if x is transitive and every element of x is also transitive.

Theorem 5.1.7. α *is an ordinal if and only if it is hereditarily transitive.*

Proof. Suppose that α is hereditarily transitive but not an ordinal. Then α is not totally ordered by $\in$. Let $x \mid y$ denote that x and y are incomparable. Let z be minimal in $\{x \in \alpha : (\exists y < \alpha)\, x \mid y\}$. Now, let y be minimal among the elements of α which are not comparable to z. We obtain the contradiction $y = z$. Given $u \in y$, the minimality of y implies that u is comparable to z. If $u = z$, we have the contradiction $z \in y$. If $z \in u$, then the transitivity of y again gives us $z \in y$. Thus, we must have $u \in z$. Since u was arbitrary, it follows that $y \subseteq z$. Next, let $u \in z$. The minimality of z now implies that u is comparable to y, and as above, it follows that in fact $u \in z$. So, $z \subseteq y$, and thus by Extensionality, $y = z$. $\qquad\square$

What kind of ordinals are there? One infinite ordinal is ω. If α is an ordinal, one can form its *successor*, the ordinal $s(\alpha) = \alpha \cup \{\alpha\}$. Putting together ω with all the ordinals obtained from ω by iterating the successor operation infinitely many times, we obtain the first *limit* ordinal, $\omega + \omega$, larger than ω. The process can then be repeated, yielding larger and larger ordinals. In general, we define the following:

Definition 5.1.8. An ordinal α is a *successor ordinal* if there is a largest ordinal β strictly smaller than α. In this case, write $\alpha = \beta + 1$. If α is not a successor ordinal or zero, then it is a *limit ordinal*.

Exercises for Section 5.1

Exercise 5.1.1. Show that for any two ordinals $\alpha \neq \beta$, $\alpha \in \beta$ if and only if $\alpha \subseteq \beta$.

Exercise 5.1.2. Prove that if α is an ordinal, then $s(\alpha)$ is an ordinal.

Exercise 5.1.3. Show that $s(\alpha)$ is an ordinal and is the the least ordinal greater than α.

Exercise 5.1.4. Show that α is a successor ordinal if and only if $\alpha = s(\beta)$ for some ordinal β.

Exercise 5.1.5. Show that an ordinal $\alpha \neq 0$ is a limit ordinal if and only if, for any ordinal $\beta < \alpha$, $s(\beta) < \alpha$.

Exercise 5.1.6. Verify that every natural number as well as ω is an ordinal.

Exercise 5.1.7. Show that for any two ordinals $\alpha \neq \beta$, $\alpha \in \beta$ if and only if $\alpha \subseteq \beta$.

Exercise 5.1.8. Show that for any two ordinals α and β, $\alpha \in \beta$ if and only if $s(\alpha) \in \beta$ or $s(\alpha) = \beta$.

Exercise 5.1.9. Show that for any non-empty class A of ordinals, $\bigcap A$ is an ordinal.

Exercise 5.1.10. Show that for any ordinal α, either $\alpha = \bigcup \alpha$ or $\alpha = s(\bigcup \alpha)$.

Exercise 5.1.11. Prove that for any set A of ordinals, $\bigcup A$ is an ordinal.

Exercise 5.1.12. Prove that for every ordinal α, there is a limit ordinal β such that $\alpha < \beta$.

Exercise 5.1.13. Prove Corollary 5.1.4 without the use of the Axiom of Regularity. *Hint.* Apply a Russell's paradox type of reasoning to the "set" of all ordinals.

5.2 Transfinite Induction and Recursion

The ordinal numbers allow proofs by transfinite induction and definitions by transfinite recursion much like natural numbers allow proofs by induction and definitions by recursion.

Theorem 5.2.1 (Transfinite Induction). *Suppose that ϕ is a formula of set theory with parameters. Suppose that $\phi(0)$ holds, and for* *every ordinal α, $[(\forall \beta \in \alpha)\,\phi(\beta)] \to \phi(\alpha)$ holds. Then, for every ordinal α, $\phi(\alpha)$ holds.*

Proof. Suppose for contradiction that there is an ordinal, call it γ, such that $\phi(\gamma)$ fails. Consider the set $x = \{\alpha \in \gamma + 1 : \neg\phi(\alpha)\}$. This is a non-empty set of ordinals, containing at least γ itself. By the Axiom of Regularity, the set x has an $\in$-minimal element α. Then $\forall \beta \in \alpha\ \phi(\beta)$ holds and $\phi(\alpha)$ fails, contradicting the assumptions. □

As in the case of induction on natural numbers, we refer to the implication $[(\forall \beta \in \alpha)\,\phi(\beta)] \to \phi(\alpha)$ as the *induction step*. In most transfinite induction arguments, the proof of induction step is divided into the successor case and the limit case according to whether α is a successor or a limit ordinal. This is actually a special case of the more general induction on sets.

Theorem 5.2.2 (Set Induction). *Suppose that ϕ is a formula of set theory with parameters. Suppose that, for every set x, $[(\forall y \in x)\,\phi(y)] \to \phi(x)$. Then $\phi(x)$ holds for every set x.*

Proof. Suppose for a contradiction that $\neg\phi(A)$ for some set A. Let $S = \{x \in \mathtt{trcl}(A) : \neg\phi(x)\}$. S is non-empty since $A \in S$ and therefore contains a minimal element x by Regularity. For any element $y \in x$, $y \in \mathtt{trcl}(A)$ since $y \in x \in A$, and therefore $\phi(y)$. It follows from the hypothesis that $\phi(x)$, which contradicts $x \in S$. □

Theorem 5.2.3 (Transfinite Recursive Definitions). *Suppose that F is a class function such that $F(x, y)$ is defined for all sets x and y. Then there is a unique class function G such that for all ordinals α and all sets x, $G(x, \alpha) = F(x, G_x \restriction \alpha)$, where $G_x \restriction \alpha = \{\langle \beta, G(x, \beta)\rangle : \beta < \alpha\}$.*

Proof. Let a given set x be fixed. We prove that for every ordinal β, there is a unique set function $G_{x,\beta}$ such that

$(*)$ $Dmn(G_{x,\beta}) = \beta$ and for every ordinal $\alpha \in \beta$, $G_{x,\beta}(\alpha) = F(x, G_{x,\beta} \restriction \alpha)$.

The proof is by induction on β. Suppose the statement is true for all $\gamma < \beta$. There are three cases to consider:

Base Step. $\beta = \emptyset$. Then $G_{x,\beta} = \emptyset$.

Successor Step. β is a successor ordinal, $\beta = \gamma+1$. Then by induction, $G_{x,\gamma}$ is a set function with domain γ and, for every ordinal $\alpha \in \gamma$, $G_{x,\gamma}(\alpha) = F(x, G_{x,\gamma} \restriction \alpha)$.

Then we have $G_{x,\beta} = G_{x,\gamma} \cup \{(\gamma, G_{x,\gamma})\}$ as the unique set function satisfying $(*)$.

Limit Step. β is a limit ordinal. Then by induction, for any $\gamma < \beta$, there is a unique set function $G_{x,\gamma}$ with domain γ such that

$$(\forall \alpha < \gamma)[G_{x,\gamma}(\alpha) = F(x, G_{x,\gamma} \restriction \alpha)].$$

Then we may define the unique function $G_{x,\beta}$ with domain β as follows:

$$G_{x,\beta}(\gamma) = z \iff (\exists g)[(\forall \alpha \leq \gamma)[g(\alpha) = F(x, g \restriction \alpha)] \wedge g(\gamma) = z].$$

For a given $\gamma < \beta$, the desired function g is $G_{x,\gamma+1}$.

Finally, the desired class function G may be defined as in the limit step above by

$$G(x,\gamma) = z \iff (\exists g)[(\forall \alpha \leq \gamma)g(\alpha) = F(x, g \restriction \alpha \wedge g(\gamma) = z].$$

The desired function g is the function $G_{x,\gamma+1}$ which we have shown exists for all ordinals γ. □

Note that a function G of one ordinal variable may be defined by transfinite recursion using $G(\alpha) = F(G \restriction \alpha)$, simply by introducing an inert variable x. That is, let $F'(x,y) = F(y)$ for all x, defining $G'(x,\alpha) = G(\alpha)$.

Next, we apply this result to monotone class operators. If Γ is not finitary, then Γ^ω need not be a fixed point, so we can continue to define Γ^α for all ordinals by letting $\Gamma^{\alpha+1} = \Gamma(\Gamma^\alpha)$ and $\Gamma^\lambda = \bigcup_{\alpha < \lambda} \Gamma^\alpha$.

Proposition 5.2.4. *For any monotone class operator Γ and any ordinal $\alpha \leq \beta$,*

1. $\Gamma^\alpha \subseteq \Gamma^{\alpha+1}$;
2. $\Gamma^\alpha \subseteq \Gamma^\beta$;
3. $\Gamma^\alpha = \bigcup_{\gamma < \alpha} \Gamma^{\gamma+1}$.

Proof. We prove parts (1) and (2) and leave (3) as an exercise. The proofs are by induction. Here is the proof of part (1).

Base Step. $\Gamma^0 = \emptyset$, so $\Gamma^0 \subseteq \Gamma^1$.

Successor Step. α is a successor ordinal, and $\alpha = \gamma + 1$ for some ordinal γ. Then by induction, $\Gamma^\gamma \subseteq \Gamma^{\gamma+1}$. Since Γ is monotone, $\Gamma^{\gamma+1} = \Gamma(\Gamma^\gamma) \subseteq \Gamma(\Gamma^{\gamma+1}) = \Gamma^{\gamma+2}$. Since $\alpha = \gamma + 1$, this proves that $\Gamma^\alpha \subseteq \Gamma^{\alpha+1}$.

Limit Step. α is a limit ordinal. By induction, we may assume that, for any $\gamma < \alpha$, $\Gamma^\gamma \subseteq \Gamma^{\gamma+1}$. Furthermore, $\Gamma^\alpha = \bigcup_{\gamma<\alpha} \Gamma^\gamma$. Now, suppose that $x \in \Gamma^\alpha$. Then $x \in \Gamma^\gamma$ for some $\gamma < \alpha$. By the assumption, $x \in \Gamma^{\gamma+1}$. Since Γ is monotone and $\Gamma^\gamma \subseteq \Gamma^\alpha$, it follows that $\Gamma^{\gamma+1} \subseteq \Gamma^{\alpha+1}$. Thus, $x \in \Gamma^{\alpha+1}$. Since x was arbitrary, this shows that $\Gamma^\alpha \subseteq \Gamma^{\alpha+1}$, as desired.

The proof of part (2) is by induction on α.

Base Step. Since $\Gamma^0 = \emptyset$, we have $\Gamma^0 \subseteq \Gamma^\beta$ for all β.

Successor Step. Assume the statement holds for α and let $s(\alpha) < \beta$. There are two subcases to consider, depending on whether β is itself a limit or a successor.

Subcase 1. β is a successor so that $\beta = \gamma + 1$. Then by induction, $\Gamma^\alpha \subseteq \Gamma^\gamma$ and it follows by monotonicity that $\Gamma^{\alpha+1} \subseteq \Gamma^{\gamma+1} = \Gamma^\beta$.

Subcase 2. β is a limit. Then $\alpha < \alpha + 1 < \beta$. Then by definition, $\Gamma^{\alpha+1} \subseteq \bigcup_{\gamma<\beta} \Gamma^\gamma = \Gamma^\beta$.

Limit Step. α is a limit. By induction, the statement holds for all $\gamma < \alpha$. Let $\alpha < \beta$. Then clearly $\gamma < \beta$ for all $\gamma < \alpha$. Thus, by induction, $\Gamma^\gamma \subseteq \Gamma^\beta$ for all $\gamma < \alpha$. Thus, $\Gamma^\alpha = \bigcup_{\gamma<\alpha} \Gamma^\gamma \subseteq \Gamma^\beta$. $\square$

Theorem 5.2.5. *If Γ is a monotone class operator, then the function $G(\alpha) = \Gamma^\alpha$ is a class function.*

Proof. Let $G(\alpha) = \Gamma^\alpha$. By Lemma 5.2.4, $G(\alpha) = \bigcup_{\beta<\alpha} \Gamma(G(\beta))$ for any α. Thus, the function G can be defined by the transfinite recursion $G(\alpha) = F(G \restriction \alpha)$, where $F(g) = \bigcup_{\beta \in Dmn(g)} \Gamma(g(\beta))$. $\square$

We can now extend the hierarchy of V_n defined previously for $n \le \omega$ to the transfinite by using the operator $\Gamma(X) = \mathcal{P}(X)$ as before. It follows that $V_{\alpha+1} = \mathcal{P}(V_\alpha)$ for all ordinals α and that and $V_\lambda = \bigcup_{\beta<\lambda} V_\beta$ if λ is a limit.

Corollary 5.2.6. *For any ordinals $\alpha \leq \beta$, $V_\alpha \subseteq V_\beta$.*

Theorem 5.2.7. *For any ordinal α, V_α is transitive.*

Proof. This is proved by induction on α. For $\alpha \leq \omega$, this is already known from Proposition 4.3.26.

Successor Step. Suppose that V_α is transitive and that $x \in y \in V_{\alpha+1}$. Then $y \in \mathcal{P}(V_\alpha)$ so that $y \subseteq V_\alpha$ and therefore $x \in V_\alpha$. It now follows from Corollary 5.2.6 that $x \in V_{\alpha+1}$. Thus, $V_{\alpha+1}$ is transitive.

Limit Step. For the case of a limit ordinal λ, suppose that $x \in y \in V_\lambda$. Then $x \in y \in V_\alpha$ for some $\alpha < \lambda$. By induction, V_α is transitive so that $x \in V_\alpha$ and hence $x \in V_\lambda$. Thus, V_λ is transitive. $\square$

Theorem 5.2.8. *For every set x, there is an ordinal α such that $x \in V_\alpha$.*

Proof. The proof uses the Axiom of Regularity. Let $V = \bigcup_\alpha V_\alpha$. Suppose that the complement of V is a non-empty class. By the Axiom of Regularity for classes (Corollary 4.3.25) applied to the complement of V, there is a set $x \notin V$ such that all its elements are in V. For every $y \in x$, let $\mathrm{rk}(y)$ be the least ordinal α such that $y \in V_{\alpha+1}$; this exists as $y \in V$ by the minimal choice of x. Now, let $F(y) = \mathrm{rk}(y) + 1$ for $y \in x$ so that $y \in V_{F(y)}$. By the Axiom of Replacement, $F[x] \subseteq Ord$ is a set. By Exercise 5.1.11, $\bigcup F[x] = \alpha$ is an ordinal, and it follows from Corollary 5.2.6 that $x \subseteq V_\alpha$. Thus, $x \in V_{\alpha+1}$ by the definition of $V_{\alpha+1}$. This contradicts the assumption that $x \notin V$. $\square$

The theorem makes it possible to define, for every set x, the ordinal $\mathrm{rk}(x)$ to be the smallest α such that $x \in V_{\alpha+1}$. The rank can serve as a rough measure of complexity of mathematical considerations. The theory of finite sets (such as most of finite combinatorics or finite group theory) takes place inside the structure $\langle V_\omega, \in \rangle$. Most mathematical analysis can be interpreted as statements about $V_{\omega+1}$. On the other hand, classical set theory often studies phenomena occurring high in the cumulative hierarchy. The high and low stages of the hierarchy are tied together more closely than one might expect.

Exercises for Section 5.2

Exercise 5.2.1. Show that, for any set x, $rk(x)$ is the least ordinal α such that $x \subseteq V_\alpha$.

Exercise 5.2.2. Show that for any sets x and y and any ordinal α,

1. if x, y are both in V_α, then $\{x, y\} \in V_{\alpha+1}$;
2. $\mathbf{rk}(\{x, y\}) = \max\{\mathbf{rk}(x), \mathbf{rk}(y)\} + 1$.

Exercise 5.2.3. Show that for any sets x and y and any ordinal α,

1. if x, y are both in V_α, then $x \cup y \in V_\alpha$;
2. $\mathbf{rk}(x \cup y) = \max\{\mathbf{rk}(x), \mathbf{rk}(y)\}$.

Exercise 5.2.4. Show that for any set x and any ordinal α,

1. if $x \in V_\alpha$, then $\mathcal{P}(x) \in V_{\alpha+1}$;
2. $\mathbf{rk}(\mathcal{P}(x)) = \mathbf{rk}(x) + 1$.

Exercise 5.2.5. Prove that for each ordinal α, $\mathbf{rk}(\alpha) = \alpha$.

5.3 Ordinal Arithmetic

In this section, we define ordinal addition, multiplication, and exponentiation by transfinite recursion, analogous to those operations on natural numbers. We derive several properties of these ordinal operations.

Our notion of transfinite recursive definitions in Theorem 5.2.3 is a course-of-values definition. For ordinal arithmetic, we want to define a function G in cases so that $G(x, 0) = H(x)$, $G(x, \alpha + 1) = J(x, G(x, \alpha))$, and $G(x, \lambda) = F(x, G_x \upharpoonright \lambda)$ for limit ordinals λ. We need to verify that this fits under Theorem 5.2.3.

Theorem 5.3.1. *Suppose that F, H, and J are three class functions which are defined for all sets. Then there is a class function G such that, for all x, all ordinals α, and all limit ordinals λ, $G(x, 0) = H(x)$, $G(x, \alpha + 1) = J(x, G(x, \alpha))$, and $G(x, \lambda) = F(x, G_x \upharpoonright \lambda)$, where $G_x(\alpha) = G(x, \alpha)$.*

Proof. Define the class function $\hat{F}$ so that $G(x, \alpha) = \hat{F}(x, G_x \upharpoonright \alpha)$ for all x and all α, by cases as follows, where α is the least ordinal not in the domain of g:

$$\hat{F}(x,g) = \begin{cases} H(x), & \text{if } \alpha = 0, \\ J(x, g(\beta)), & \text{if } \alpha = \beta + 1, \\ F(x, g), & \text{if } \alpha \text{ is a limit ordinal.} \end{cases}$$

The existence of the class function G now follows from Theorem 5.2.3. □

Here is the definition of ordinal addition.

Definition 5.3.2. For all ordinals α and β and for all limit ordinals λ,

1. $\alpha + 0 = \alpha$;
2. $\alpha + s(\beta) = s(\alpha + \beta)$;
3. $\alpha + \lambda = \bigcup_{\beta < \lambda} (\alpha + \beta)$.

This seems like a straightforward generalization of addition for natural numbers, and there are many similarities. However, there are also a number of surprising differences.

For example, $2 + \omega = \bigcup_{n \in \omega} 2 + n = \omega$, since each $2 + n \in \omega$. But $\omega \in s(s(\omega)) = \omega + 2$. Thus, $2 + \omega \neq \omega + 2$. Thus, ordinal addition is not in general commutative.

Lemma 5.3.3.

1. *For all α, β and γ, $\beta < \gamma$ implies $\alpha + \beta < \alpha + \gamma$.*
2. *For all α, β and γ, $\beta < \gamma$ implies $\beta + \alpha \leq \gamma + \alpha$.*

Proof. (1) The proof is by induction on γ.

Base Step. For $\gamma = 0$, this is vacuous.

Successor Step. For the successor case, suppose that $\gamma = s(\delta)$ for some ordinal δ. Then either $\beta < \delta$ or $\beta = \delta$. If $\beta = \delta$, then $\alpha + \beta = \alpha + \delta < s(\alpha + \delta) = \alpha + s(\delta) = \alpha + \gamma$. If $\beta < \delta$, then by induction, $\alpha + \beta < \alpha + \delta$ so that again $\alpha + \beta < \alpha + s(\delta) = \alpha + \gamma$.

Limit Step. The limit case does not actually require induction. If γ is a limit and $\beta < \gamma$ so that $s(\beta) < \gamma$, then $\alpha + \beta < \alpha + s(\beta)$ and $\alpha + s(\beta) \subseteq \alpha + \gamma$ since $\alpha + \gamma = \bigcup_{\delta < \gamma} \alpha + \delta$ by definition.

(2) is left as an exercise. □

Lemma 5.3.4 (Subtraction Lemma). *For all ordinals $\alpha \leq \beta$, there exists a unique δ such that $\beta = \alpha + \delta$.*

Proof. Let δ be the least such that $\alpha + \delta \geq \beta$. This exists since $\alpha + \beta \geq 0 + \beta = \beta$. It suffices to show that $\alpha + \delta \leq \beta$. Suppose by way of contradiction that $\alpha + \delta > \beta$. There are two cases: If $\delta = s(\gamma)$ for some γ and $\alpha + \delta > \beta$, then $\gamma < \delta$, but $\alpha + \gamma \geq \beta$, which violates the choice of δ as the least. If δ is a limit and $\alpha + \delta > \beta$, then there must exist $\gamma < \delta$ such that $\alpha + \gamma \geq \beta$, again violating the choice of δ as the least.

For uniqueness, suppose that there is δ' such that $\alpha + \delta' = \alpha + \delta = \beta$. If $\delta' < \delta$, then this contradicts the choice of δ as the least ordinal satisfying $\alpha + \delta \geq \beta$. If, however, $\delta < \delta'$, then by Lemma 5.3.3, $\alpha + \delta < \alpha + \delta'$, which is impossible. Thus, $\delta = \delta'$. $\square$

Every infinite countable ordinal α is bijective with ω, and this bijection induces a well-ordering of ω which is isomorphic to $(\alpha, \in)$. For example, define an ordering R of ω such that $0R1$, and, for all $x, y > 1$, we have $xR0$, $xR1$, and $xRy \iff x < y$. Then (ω, R) is isomorphic to $\omega + 2$.

Recall that for two ordered sets $(A, \leq_A)$ and $(B, \leq_B)$, we defined an order $\leq$ on the disjoint union $A \oplus B$ which put a copy of B after a copy of A.

Proposition 5.3.5. *For any ordinals α and β, $(\alpha, \in) \oplus (\beta, \in)$ is isomorphic to $(\alpha + \beta, \in)$.*

Proof. Define the mapping $f : \alpha \oplus \beta$ to $\alpha + \beta$ by letting $f(0, x) = x$ and $f(1, x) = \alpha + x$. f is order-preserving by Lemma 5.3.3. f is onto by the Subtraction Lemma (Lemma 5.3.4). $\square$

Next, we consider ordinal multiplication.

Definition 5.3.6. For all ordinals α and β and for all limit ordinals λ,

1. $\alpha \cdot 0 = 0$;
2. $\alpha \cdot s(\beta) = (\alpha \cdot \beta) + \alpha$;
3. $\alpha \cdot \lambda = \bigcup_{\beta < \lambda} (\alpha \cdot \beta)$.

Again this seems like a straightforward generalization of multiplication for natural numbers, and there are many similarities. However, there are also a number of differences.

For example, $2 \cdot \omega = \bigcup_{n \in \omega} 2n = \omega$, since each $2n \in \omega$. But $\omega \cdot 2 = \omega + \omega > \omega$. In addition, the example

$$(\omega + 1) \cdot 2 = \omega + 1 + \omega + 1 = \omega + \omega + 1 \neq \omega + \omega + 2 = \omega \cdot 2 + 1 \cdot 2$$

shows that the distributive law fails on one side. It is an exercise to show that it does hold on the other side.

Lemma 5.3.7.

1. *For all α, β and γ, $\beta < \gamma$ implies $\alpha \cdot \beta < \alpha \cdot \gamma$.*
2. *For all α, β and γ, $\beta < \gamma$ implies $\beta \cdot \alpha \leq \gamma \cdot \alpha$.*

The proofs are left to the exercises.

Lemma 5.3.8 (Division Lemma). *For all ordinals α and β, there exist unique ordinals $\rho < \alpha$ and δ such that $\beta = \alpha \cdot \delta + \rho$.*

Proof. Let δ be the least ordinal such that $\alpha \cdot s(\delta) > \beta$. It follows that $\alpha \cdot \delta \leq \beta$. This can be seen by considering two cases: If $\delta = s(\gamma)$ and $\alpha \cdot \delta > \beta$, then $\gamma < \delta$ and this violates the choice of δ as the least. If δ is a limit and $\alpha \cdot \delta > \beta$, then there must exist $\gamma < \delta$ such that $\alpha \cdot s(\gamma) > \beta$, again violating the choice of δ as the least.

Now, let ρ be given by the Subtraction Lemma so that $\beta = \alpha \cdot \delta + \rho$. The uniqueness of ρ follows from the Subtraction Lemma. We leave the proof of the uniqueness of δ as an exercise. $\square$

If $\mathcal{A} = (A, <_A)$ and $\mathcal{B} = (B, <_B)$ are two linear orderings, we may define the product $\mathcal{A} \otimes \mathcal{B}$ of these orderings to have universe $A \times B$ and ordering defined, for all $a_1, a_2 \in A$ and all $b_1, b_2 \in B$, by

$$\langle a_1, b_1 \rangle < \langle a_2, b_2 \rangle \iff a_1 <_A a_2 \lor (a_1 = a_2 \land b_1 <_B b_2).$$

Proposition 5.3.9. *For any ordinals α and β, $(\beta, \in) \times (\alpha, \in)$ is isomorphic to $(\alpha \cdot \beta, \in)$.*

Proof. Define the mapping $f : \beta \otimes \alpha$ to $\alpha \cdot \beta$ by letting $f(y, x) = \alpha \cdot y + x$. We show the function f is order-preserving using Lemma 5.3.7. Suppose that $(y_1, x_1) \leq (y_2, x_2)$ in the ordering defined above on $\beta \otimes \alpha$. There are two cases to consider. First, we may have $y_1 < y_2$. Then since $x_1 < \alpha$,

$$f(y_1, x_1) = \alpha \cdot y_1 + x_1 < \alpha \cdot y_1 + \alpha$$
$$= \alpha \cdot s(y_1) \leq \alpha \cdot y_2 \leq \alpha \cdot y_2 + x_2 = f(y_2, x_2).$$

Second, we may have $y_1 = y_2$ and $x_1 < x_2$. Then

$$f(y_1, x_1) = \alpha \cdot y_1 + x_1 = \alpha \cdot y_2 + x_1 < \alpha \cdot y_2 + x_2 = f(y_2, x_2).$$

Lastly, the function f is onto by the Division Lemma. That is, given any ordinal γ, there exist $x < \alpha$ and y such that $\gamma = \alpha \cdot y + x$. Given $\gamma < \alpha \cdot \beta$, it follows that the above y from the Division Lemma satisfies $y < \beta$ and then we have $f(y, x) = \gamma$. $\qquad \square$

Finally, we consider ordinal exponentiation.

Definition 5.3.10. For all ordinals $\alpha > 0$ and β and for all limit ordinals λ,

1. $\alpha^0 = 1$;
2. $\alpha^{s(\beta)} = \alpha^\beta \cdot \alpha$;
3. $\alpha^\lambda = \bigcup_{\beta < \lambda} \alpha^\beta$.

This is a natural generalization of multiplication for natural numbers, and there are many similarities. Once again, there are also a number of differences.

For example, $2^\omega = \bigcup_{n \in \omega} 2^n = \omega$, since each $2^n \in \omega$. So, the cardinality of 2^ω as obtained by ordinal exponentiation is not the same as the cardinality of the set $\{0, 1\}^\omega$ of functions mapping ω into $\{0, 1\}$.

It is easy to see that $1^\beta = 1$ for all ordinals β and that α^λ is a limit ordinal for any $\alpha \geq 2$ and any limit ordinal λ. These are left as exercises.

Lemma 5.3.11.

1. *For all $\alpha \geq 2$, all β and γ, $\beta < \gamma$ implies $\alpha^\beta < \alpha^\gamma$.*
2. *For all α, all $\beta > 0$ and γ, $\beta < \gamma$ implies $\beta^\alpha \leq \gamma^\alpha$.*

The proof of Lemma 5.3.11 is left to the exercises. Two important identities from the exponentiation of real numbers carry over to ordinal exponentiation. We first prove a general lemma to simplify the argument.

Lemma 5.3.12. *Suppose that λ a limit ordinal and $f : \beta \to \lambda$ is an ordinal function such that $\lambda = \bigcup_{\gamma < \beta} f(\gamma)$. Then for any ordinal α,*

1. $\alpha + \lambda = \bigcup_{\gamma < \beta} \alpha + f(\gamma)$;
2. $\alpha \cdot \lambda = \bigcup_{\gamma < \beta} \alpha \cdot f(\gamma)$;
3. $\alpha^\lambda = \bigcup_{\gamma < \beta} \alpha^{f(\gamma)}$.

Proof. Let $A = \alpha + \lambda$. By definition, $A = \bigcup_{\delta < \lambda} \alpha + \delta$. Let $B = \bigcup_{\gamma < \beta} \alpha + f(\gamma)$. We show that these are equal.

Suppose first that $x \in A$. Then $x \in \alpha + \delta$ for some $\delta < \lambda$. Since $\lambda = \bigcup_{\gamma < \beta} f(\gamma)$, there is some $\gamma < \beta$ such that $\delta < f(\gamma)$ and hence $\alpha + \delta < \alpha + f(\gamma)$ so that $x \in B$.

Suppose next that $x \in B$. Then $x \in \alpha + f(\gamma)$ for some $\gamma < \beta$. Then $\delta = f(\gamma) < \lambda$, and this shows that $x \in A$.

The proofs for multiplication and exponentiation are similar. □

This result can be used to demonstrate the associative law for ordinal addition and multiplication. The following lemma is needed.

Lemma 5.3.13. *For any ordinal β and any limit ordinal λ, $\beta + \lambda$ is a limit ordinal.*

Proof. Let $\delta \in \beta + \lambda$. Then by definition, $\delta \in \beta + \gamma$ for some $\gamma < \lambda$. Then $s(\delta) \in \beta + s(\gamma)$ by Lemma 5.3.3, since for ordinals $x \in y \iff x < y$. Since λ is a limit, $s(\gamma) < \lambda$ and therefore $s(\delta) \in \beta + \lambda$ as well. □

Lemma 5.3.14. *For all α, β, and γ,*

1. $(\alpha + \beta) + \gamma = \alpha + (\beta + \gamma)$;
2. $(\alpha \cdot \beta) \cdot \gamma = \alpha \cdot (\beta \cdot \gamma)$.

Proof. (1) The proof is by induction on γ.

Base Step. For $\gamma = 0$, $(\alpha + \beta) + 0 = \alpha + \beta = \alpha + (\beta + 0)$.

Successor Step. For the successor case, suppose that $\gamma = s(\delta)$ for some ordinal δ. Then by induction, $(\alpha + \beta) + \delta = \alpha + (\beta + \delta)$. Thus, we have

$$(\alpha + \beta) + s(\delta) = s((\alpha + \beta) + \delta)$$
$$= s(\alpha + (\beta + \delta)) = \alpha + s(\beta + \delta) = \alpha + (\beta + s(\delta)).$$

Limit Step. Suppose that λ is a limit ordinal and that $(\alpha + \beta) + \gamma = \alpha + (\beta + \gamma)$ for all $\gamma < \lambda$. Since $\beta + \lambda = \bigcup_{\gamma < \lambda} \beta + \gamma$, it follows from Lemma 5.3.12 (with $f(\gamma) = \beta + \gamma$) that

$$\alpha + (\beta + \lambda) = \bigcup_{\gamma < \lambda} \alpha + (\beta + \gamma) = \bigcup_{\gamma < \lambda} (\alpha + \beta) + \gamma = (\alpha + \beta) + \lambda.$$

The proof of part (2) is left as an exercise. □

Proposition 5.3.15. *For all $\alpha \neq 0$, $\alpha^{\beta+\gamma} = \alpha^\beta \cdot \alpha^\gamma$.*

Proof. The proof is by induction on γ.

Base Step. For $\gamma = 0$, we have $\alpha^{\beta+0} = \alpha^\beta = \alpha^\beta \cdot \alpha^0$.

Successor Step. Given $\gamma = \delta + 1$ for some δ, we have $\alpha^{\beta+\delta} = \alpha^\beta \cdot \alpha^\delta$. Then

$$\alpha^{\beta+\gamma} = \alpha^{\beta+s(\delta)} = \alpha^{s(\beta+\delta)} = \alpha^{\beta+\delta} \cdot \alpha = (\alpha^\beta \cdot \alpha^\delta) \cdot \alpha$$
$$= \alpha^\beta \cdot (\alpha^\delta \cdot \alpha) = \alpha^\beta \cdot \alpha^{s(\delta)} = \alpha^\beta \cdot \alpha^\gamma.$$

Limit Step. If γ is a limit, suppose that $\alpha^{\beta+\delta} = \alpha^\beta \cdot \alpha^\delta$ for all $\delta < \gamma$. Then

$$\alpha^{\beta+\gamma} = \bigcup_{\delta < \gamma} \alpha^{\beta+\delta} = \bigcup_{\delta < \gamma} \alpha^\beta \cdot \alpha^\delta.$$

But this equals $\alpha^\beta \cdot \alpha^\gamma$ by (2) of Lemma 5.3.12, since α^γ is a limit ordinal and $\alpha^\gamma = \bigcup_{\delta < \gamma} \alpha^\delta$. $\square$

Proposition 5.3.16. *For all $\alpha \neq 0$, $\alpha^{\beta \cdot \gamma} = (\alpha^\beta)^\gamma$.*

The proof is left as an exercise.

Similar to the Subtraction Lemma and the Division Lemma, we have the following.

Lemma 5.3.17 (Logarithm Lemma). *For all ordinals $\alpha > 0$ and $\beta > 1$, there exist unique ordinals γ, ρ, and δ such that $\alpha = \beta^\gamma \cdot \delta + \rho$ and $\gamma \leq \alpha$, $0 < \delta < \beta$, and $\rho < \beta^\gamma$.*

Proof. Let γ be the least ordinal such that $\beta^{s(\gamma)} > \alpha$. It follows that $\beta^\gamma \leq \alpha$. This can be seen by considering two cases. If $\gamma = s(\tau)$ is a successor and $\beta^{s(\tau)} > \alpha$, then $\tau < \gamma$ violates the choice of γ as the least such ordinal. If γ is a limit and $\beta^\gamma > \alpha$, then there must exist $\tau < \gamma$ such that $\beta^{s(\tau)} > \alpha$, again violating the choice of γ as the least such ordinal.

Now, let δ and ρ be given by the Division Lemma so that $\alpha = \beta^\gamma \cdot \delta + \rho$ and $\rho < \beta^\gamma$. It follows by preservation of order that $\delta < \beta$, since otherwise $\alpha = \beta^\gamma \cdot \delta + \rho \geq \beta^{s(\gamma)}$. The uniqueness of ρ and δ are guaranteed by the Division Lemma. We leave the proof of the uniqueness of γ as an exercise. $\square$

Recall the expression of a natural number in base 2 or in a more general base b from Proposition 4.4.5 and Exercise 4.4.2. For example, in base 3, the number 23 is given as 212, indicating that $23 = 3^2 \cdot 2 + 3^1 \cdot 1 + 3^0 \cdot 2$. We can use the Division Lemma to give a similar representation for countable ordinals using base ω.

Theorem 5.3.18 (Cantor Normal Form). *For any ordinal $\alpha > 0$, there exist unique finite sequences $\gamma_0 < \gamma_1 < \cdots < \gamma_{m-1} \leq \alpha$ and $n_0, \ldots, n_{m-1} < \omega$, with $n_{m-1} > 0$, such that*

$$\alpha = \omega^{\gamma_{m-1}} \cdot n_{m-1} + \cdots + \omega^{\gamma_0} \cdot n_0.$$

Proof. The proof is by induction on α. If α is finite, then $m = 1$, $\gamma_0 = 0$ and $n_0 = \alpha$.

Now, suppose that the theorem holds for all ordinals less than α and $\alpha \geq \omega$. Then by the Logarithm Lemma, we have $\alpha = \omega^\gamma \cdot n + \rho$ for some $n < \omega$ and some $\rho < \omega^\gamma$. By induction, we have $\gamma_0 < \gamma_1 < \cdots < \gamma_{m-1} \leq \rho$ and $n_0, \ldots, n_{m-1} < \omega$ such that $\rho = \omega^{\gamma_{m-1}} \cdot n_{m-1} + \cdots + \omega^{\gamma_0} \cdot n_0$, where $\gamma_{m-1} < \gamma$ since $\omega^{\gamma_{m-1}} \leq \rho < \omega^\gamma$. It follows that $\alpha = \omega^\gamma \cdot n + \omega^{\gamma_{m-1}} \cdot n_{m-1} + \cdots + \omega^{\gamma_0} \cdot n_0$. $\qquad\square$

Proof of the uniqueness of the sequences of ordinals and natural numbers is left to the exercises.

The following notions are of interest.

Definition 5.3.19. Let α be a limit ordinal and let $C \subseteq \alpha$.

1. C is *bounded* in α if there exists $\beta < \alpha$ such that $C \subseteq \beta$. (Otherwise C is *unbounded*.)
2. C is *closed* in α if $\bigcup A \in C$ whenever A is a non-empty bounded subset of C.

Example 5.3.20. The set $\{\omega \cdot n : n \in \omega\}$ is closed and unbounded in $\omega \cdot \omega$. The set $(\omega + \omega) \setminus \{\omega\}$ is unbounded in $\omega + \omega$ but is not closed since $\bigcup \omega = \omega$.

Definition 5.3.21. Let $F : Ord \to Ord$ be a class functional.

1. F is *strictly increasing* if, for any ordinals α and β, $\alpha < \beta$ implies that $F(\alpha) < F(\beta)$.
2. F is *continuous* if, for any limit ordinal α, $F(\alpha) = \bigcup \{F(\beta) : \beta < \alpha\}$.
3. F is *normal* if F is both strictly increasing and continuous.

Example 5.3.22. Fix an ordinal α and let $F(x) = \alpha + x$ for any ordinal x. It follows from Definition 5.3.2 that F is continuous and it follows from Lemma 5.3.3 that F is strictly increasing. Thus, F is normal.

Lemma 5.3.23. *Suppose that α is a limit ordinal and $F : \alpha \to \alpha$ is normal. Then for any non-empty bounded subset C of α, $\bigcup F[C] = F(\bigcup C)$.*

Proof. Let $\gamma = \bigcup C$. There are two cases. First, suppose that γ is a successor ordinal. Then γ is the greatest element of C so that $F(\gamma)$ is the greatest element of $F[C]$, since F is strictly increasing. It follows that $\bigcup F[C] = \gamma$.

Second, suppose that γ is a limit ordinal. Then $F(\gamma) = \bigcup F[C]$, since F is continuous. $\qquad\qquad\square$

Theorem 5.3.24. *Let C be a closed subset of a limit ordinal α, and let $F : \alpha \to \alpha$ be a normal function. Then $F[C]$ is closed.*

Proof. Assume that C is a closed subset of the limit ordinal α and $F : \alpha \to \alpha$ is normal. Let $B \subseteq F[C]$ be bounded in α. Since F is strictly increasing, it is one-to-one, and furthermore $x \le F(x)$ for all $x \in \alpha$. (See the exercises.) It follows that $B = F[A]$, where $A = F^{-1}[B]$. A must be bounded in α since B is bounded and $x \le F(x)$ for all $x \in \alpha$. Thus, $\bigcup A \in C$. Since F is continuous, $\bigcup F[A] = F(\bigcup A) \in F[C]$, and therefore $\bigcup B \in F[C]$. $\qquad\square$

We conclude the section with a brief discussion of fixed points, which are important for the study of cardinal numbers.

Definition 5.3.25. For any function f and any element x of the domain of f, we say that x is a *fixed point* of f if $f(x) = x$.

Recall that $2 + \omega = 2 \cdot \omega = 2^\omega$ in ordinal arithmetic. We can interpret this to mean that ω is a fixed point of the function $f(x) = 2 + x$, as well as the functions $2 \cdot x$ and 2^x. In fact, ω is the *least fixed point* of $2 + x$ and of 2^x. $2 \cdot 0 = 0$, so 0 is a fixed point of $2 \cdot x$. We may define larger ordinals as least fixed points of certain natural functions.

Proposition 5.3.26.

1. ω^2 is the least fixed point of the function $f(x) = \omega + x$.
2. ω^ω is the least fixed point of the function $f(x) = \omega \cdot x$.

Proof. We give the proof of the first part and leave the second part as an exercise. To see that w^2 is a fixed point of f, note that $w^2 = \bigcup_n w \cdot n$ so that $w + w^2 = \bigcup_n w + w \cdot n = \bigcup_n w \cdot (n+1) = w^2$.

We see that w^2 is the least fixed point as follows. Let $x < w^2$. Then by the Division Lemma, we can write $x = w \cdot m + n$ for some $m, n \in w$. It follows that $w + x = w \cdot (m+1) + n$ so that $x < f(x)$. □

This notion may be used to define certain countable ordinals. For example, we let ϵ_0 be the least fixed point of the function $f(x) = w^x$. To see that ϵ_0 exists, just recursively define a function F with domain w by letting $F(0) = w$ and $F(n+1) = w^{F(n)}$ for each n. Then F exists by Theorem 4.3.27 and we let $\epsilon_0 = \bigcup_n F(n)$. Clearly, ϵ_0 is a limit, since $F(n) < F(n+1)$ for each n. Now, $w^{\epsilon_0} = w^{\bigcup_n F(n)} = \bigcup_n w^{F(n)} = \bigcup_n F(n+1) = \epsilon_0$. The ordinal ϵ_0 has been of great interest in computability and logic.

A similar argument shows that any normal function has a fixed point. This is left as an exercise.

Exercises for Section 5.3

Exercise 5.3.1. Show that for all β, $0 + \beta = \beta$.

Exercise 5.3.2. Show that for all α, β, and γ, $\beta < \gamma$ implies $\beta + \alpha \leq \gamma + \alpha$.

Exercise 5.3.3. Prove that for all ordinals α and β, $\alpha < \beta$ if and only if there exists $\gamma > 0$ such that $\beta = \alpha + \gamma$.

Exercise 5.3.4. Prove that for all limit ordinals λ and all $\alpha < \lambda$, $\alpha + w \leq \lambda$.

Exercise 5.3.5. Show that for all α and all limit ordinals λ, $\alpha + \lambda$ is a limit ordinal.

Exercise 5.3.6. Define an ordering on w which is isomorphic to $w \cdot 3 + 5$.

Exercise 5.3.7. Prove that ordinal multiplication is associative.

Exercise 5.3.8. Show that for all β, $0 \cdot \beta = 0$ and $1 \cdot \beta = \beta$.

Exercise 5.3.9. Show that for all $\alpha > 0$, β, and γ, $\beta < \gamma$ implies $\alpha \cdot \beta < \alpha \cdot \gamma$.

Exercise 5.3.10. Show that for all α, β, and γ, $\beta < \gamma$ implies $\beta \cdot \alpha \leq \gamma \cdot \alpha$.

Exercise 5.3.11. Finish the proof of the Division Lemma by showing the uniqueness of the ordinal δ in the statement of the result.

Exercise 5.3.12. Prove that for all α, β, γ, $\alpha \cdot (\beta + \gamma) = \alpha \cdot \beta + \alpha \cdot \gamma$.

Exercise 5.3.13. Define an ordering on ω which is isomorphic to $\omega^3 = \omega \cdot \omega \cdot \omega$.

Exercise 5.3.14. Show that for all α and all limit ordinals λ, $\alpha \cdot \lambda$ is a limit ordinal.

Exercise 5.3.15. Show that for all limit ordinals λ, there exists α such that $\lambda = \omega \cdot \alpha$.

Exercise 5.3.16. Show that for all limit ordinals λ, $2 \cdot \lambda = \lambda$.

Exercise 5.3.17. Prove that $1^\beta = 1$ for all ordinals β.

Exercise 5.3.18. Prove that α^λ is a limit ordinal for any $\alpha \geq 2$ and any limit ordinal λ. *Hint*: Use the previous exercise.

Exercise 5.3.19. Show that for all $\alpha \geq 2$, all β and γ, $\beta < \gamma$ implies $\alpha^\beta < \alpha^\gamma$.

Exercise 5.3.20. Show that for all α, all $\beta > 0$ and γ, $\beta < \gamma$ implies $\beta^\alpha \leq \gamma^\alpha$.

Exercise 5.3.21. Show that for all $\alpha \neq 0$, $\alpha^{\beta \cdot \gamma} = (\alpha^\beta)^\gamma$.

Exercise 5.3.22. Finish the proof of the Logarithm Lemma by showing the uniqueness of the ordinal γ in the statement of the result.

Exercise 5.3.23. Prove that the Cantor Normal Form is unique.

Exercise 5.3.24. Show that the set $\{\omega^n : n \in \omega\}$ is closed and unbounded in ω^ω.

Exercise 5.3.25. Define the class function F by letting $F(x) = \omega \cdot x$ for any ordinal x. Show that F is normal.

Exercise 5.3.26. Define the class function G by letting $G(x) = x + \omega \cdot x$ for any ordinal x. Show that F is strictly increasing but is not continuous.

Exercise 5.3.27. Show that if a function $F : Ord \to Ord$ is strictly increasing, then F is one-to-one and $x \leq F(x)$ for all ordinals x. Show that this also holds for $F : \alpha \to \alpha$, where α is a limit ordinal.

Exercise 5.3.28. Suppose that $F : Ord \to Ord$ and $G : Ord \to Ord$ are normal. Show that the composition $F \circ G$ is also normal.

Exercise 5.3.29. Show that ω^ω is the least fixed point of the function $f(x) = \omega \cdot x$.

Exercise 5.3.30. For any normal class function $F : Ord \to Ord$ and any ordinal α, there exist an ordinal $\beta \geq \alpha$ such that $F(\beta) = \beta$.

5.4 Ordinals and Well-Orderings

Recall that a *well-ordering* is a linear ordering $\leq$ on a set x which in addition satisfies the condition that every non-empty subset $a \subseteq x$ has a $\leq$-least element, i.e. an element u such that the conjunction $v \in a$ and $v \leq u$ implies $v = u$.

It is clear that every ordinal is a well-ordering: Every subset of an ordinal has an $\in$-minimal element by the Axiom of Regularity, and by the linearity of $\in$, this is in fact an $\in$-smallest element. The next theorem shows that up to isomorphism, the ordinals are the only well-orderings.

Theorem 5.4.1. *Every well-ordering is isomorphic to a unique ordinal.*

Proof. The uniqueness part follows from the rigidity of ordinals, i.e. Theorem 5.1.5. For the existence part, let $\leq$ be a well-ordering on a set x. By transfinite recursion define a class function G on the class of all ordinals by letting $G(\alpha)$ be the $\leq$-least element of the set $x \setminus Rng(G \restriction \alpha)$ if the latter set is non-empty, and $G(\alpha) = \mathtt{trash}$ otherwise. We show that there is an ordinal α such that $G(\alpha) = \mathtt{trash}$, and for the least such ordinal α, the function $G \restriction \alpha : \alpha \to x$ is an isomorphism of linear orders.

Suppose for contradiction that there is no ordinal α such that $G(\alpha) = \mathtt{trash}$. Then G is an injection from the proper class of all ordinals to the set x. Such an injection does not exist by

Exercise 3.7.3. This contradiction proves the existence of an ordinal α such that $G(\alpha) = \texttt{trash}$.

Now, let α be the smallest ordinal such that $G(\alpha) = \texttt{trash}$, and consider the function $G \restriction \alpha$. Its domain is equal to α. Its range must be equal to x as this is the only way that $G(\alpha) = \texttt{trash}$ can occur. To conclude the proof, it will be enough to show that $G \restriction \alpha$ preserves the ordering given by $\in$.

Suppose for contradiction that $G \restriction \alpha$ does not preserve this ordering. Then there must be ordinals $\gamma \in \beta \in \alpha$ such that $G(\beta) < G(\gamma)$. But then, $G(\beta) \notin Rng(G \restriction \beta) \supseteq Rng(G \restriction \gamma)$. Therefore, by the recursive definition of G at γ, the element $G(\beta) \in x$ or something even smaller than it should have been picked as the value of $G(\gamma)$. This is a contradiction. □

Note the use of the Axiom Schema of Replacement, via Exercise 3.7.3, in the above proof. The theorem cannot be proved without it. The development of ordinals is one of the reasons why Replacement was incorporated into ZFC.

We define the *order type* of a well-ordering to be the unique ordinal given by Theorem 5.4.1.

Here is a property of ordinals that is needed in Chapter 6.

Proposition 5.4.2. *For any ordinal γ and any subset A of γ, the order type of $(A, \in)$ is $\leq \gamma$.*

Proof. Let the ordinal γ and $A \subseteq \gamma$ be given. Let α be the order type of $(A, \in)$ and let $f : \alpha \to A$ be the canonical isomorphism as in Theorem 5.4.1. We claim that, for all $\beta < \alpha$, $\beta \leq f(\beta)$.

Base Step. For $\beta = 0$, clearly $\beta = 0 \leq f(\beta)$.

Successor Step. Suppose that $\beta \leq f(\beta)$. Since f is an order isomorphism and $\beta < \beta + 1$, it follows that $f(\beta) < f(\beta + 1)$. Thus, $\beta < f(\beta + 1)$, and hence $\beta + 1 \leq f(\beta + 1)$.

Limit Step. Let λ be a limit and suppose by induction that $\beta \leq f(\beta)$ for all $\beta < \lambda$. Since f is an order isomorphism, this implies that $\beta \leq f(\lambda)$ for all $\beta < \lambda$, and hence $\lambda \leq f(\lambda)$.

Since $f : \alpha \to A$ and $A \subseteq \gamma$, it follows that, for all $\beta < \alpha$, $\beta \leq f(\beta) < \gamma$. Thus, $\alpha \leq \gamma$. □

We can generalize the addition of two well-orderings to an infinite sum. First, given an infinite sequence $\alpha_0, \alpha_1, \ldots$ of ordinals, we can define the sum $\sum_{i<\omega} \alpha_i$ as $\bigcup_{i<\omega} \alpha_0 + \alpha_1 + \cdots + \alpha_i$. This will be an ordinal and, if each α_i is countable, the sum will also be countable.

We can also define the infinite disjoint union of a sequence $(A_0, R_0), (A_1, R_1), \ldots$ of orderings, written as $\bigoplus_{i<\omega}(A_i, R_i)$ to have universe $\bigcup_{i<\omega}\{i\} \times A_i$ and be ordered so that

1. for $i < j$, $(i, a) < (j, b)$ for all $a \in A_i$ and all $b \in A_j$ and
2. for each i and each $a, b \in A_i$, $(i, a) < (i, b)$ if and only if aR_ib.

Proposition 5.4.3. *Suppose that* $(A_0, R_0), (A_1, R_1), \ldots$ *is an infinite sequence of well-ordered sets and each* (A_i, R_i) *is isomorphic to an ordinal* α_i. *Then* $\bigoplus_i(A_i, R_i)$ *is isomorphic to* $\sum_{i<\omega} \alpha_i$.

The proof is left as an exercise.

Lemma 5.4.4. *For any countable limit ordinal* λ, *there is an infinite sequence* $\beta_0 \leq \beta_1 \leq \ldots$ *such that* $\lambda = \bigcup_i \beta_i$ *and there is an infinite sequence* $\alpha_0, \alpha_1, \ldots$ *such that* $\lambda = \sum_i \alpha_i$.

Proof. Since λ is countable, there is a bijection $f : \omega \to \lambda$. It follows that $\lambda = \bigcup_i f(i)$. For each i, let $\beta_i = \bigcup_{n<i} f(n)$ so that $\lambda = \bigcup_i \beta_i$ as desired. Now, use the subtraction lemma to get α_i such that $\alpha_0 = \beta_0$ and, for all i, $\beta_{i+1} = \beta_i + \alpha_{i+1}$. It follows that $\lambda = \sum_i \alpha_i$. $\qquad\square$

Theorem 5.4.5. *For any countable ordinal* α, *there is a subset* P *of* $\mathbb{Q}$ *such that* α *is isomorphic to* $(P, <)$ *under the standard ordering of the rationals.*

Proof. We begin with two observations. First, $(\mathbb{Q}, <)$ is order isomorphic to $\mathbb{Q} \cap (-1, 1)$ via the mapping $f(x) = \frac{x}{1+|x|}$. Second, it is sufficient to show that there is an embedding from α into $(\mathbb{Q}, <)$ since then α would be isomorphic to the image of this embedding. We proceed by transfinite induction.

Base Step. For $\alpha = 0$, let $P = \emptyset$.

Successor Step. Let α be isomorphic to $(P, <)$, where $P \subset \mathbb{Q} \cap (-1, 1)$. Then $\alpha + 1$ is isomorphic to $P \cup \{1\}$.

Limit Step. Let λ be a countable limit ordinal and assume by induction that for each $\alpha < \lambda$, there is an embedding of α into $\mathbb{Q}$. By Lemma 5.4.4, we have a sequence $\alpha_0, \alpha_1, \ldots$ with each $\alpha_i < \lambda$ such that $\lambda = \sum_{i<\omega} \alpha_i$. By induction, there are subsets P_i of $\mathbb{Q}$ isomorphic to α_i for each i. We may assume without loss of generality that $P_i \subseteq (i, i+2) \cap \mathbb{Q}$ for each i. It follows that λ is isomorphic to $\oplus_i P_i$ which is in fact isomorphic to $\bigcup_i P_i$ since the sets P_i are pairwise disjoint and already ordered with P_i preceding P_j for each $i < j$. $\qquad\square$

Example 5.4.6. Let $P_0 = \{n + 1 - 2^{-k} : n, k \in \omega\} = \{0, \frac{1}{2}, \frac{3}{4}, \ldots, 1, \frac{3}{2}, \frac{7}{4}, \ldots, 2, \ldots\}$. This has order type ω^2. Observe that this is a closed subset of the real numbers. Consider the following two subsets of P_0. $P_1 = \{x \in P_0 : x \geq 1\}$ is closed and has order type ω^2. $P_2 = \{2n + 1 - 2^{-k} : n, k \in \omega\}$ also has order type ω^2 but is not closed, since it contains every $1 - 2^{-k}$ but does not contain 1.

It is natural to consider a connection between well-ordered closed sets of real numbers and closed sets of ordinals.

Proposition 5.4.7. *Let S be a well-ordered subset of $\mathbb{R}$ and let Φ map S to its order type α. Then P is a closed subset of S if and only if $\Phi[P]$ is a closed subset of α.*

Proof. Let P be a subset of the well-ordered set S of real numbers and let $\Phi : S \to \alpha$ map S to its order type α. Suppose first that P is closed and let $\beta_1, \beta_2, \ldots$ be an increasing sequence of ordinals in $\Phi[P]$ which converges to an ordinal $\beta < \alpha$. Let x_i be chosen, for $i \in \omega$, so that $\Phi(x_i) = \beta_i$, and let $\Phi(y) = \beta$. Then $(x_i)_{i\in\omega}$ is an increasing sequence and it follows that $y = \lim_{i\to\infty} x_i$. Since P is closed, $y \in P$ and therefore $\beta = \Phi(y) \in \Phi[P]$.

For the converse, suppose that $\Phi[P]$ is closed and let $x_1, x_2, \ldots$ be an increasing sequence of elements of P, bounded by some $z \in P$. Then $\lim_{i\in\omega} x_i = y$ for some $y \leq z$. Let $\beta_i = \Phi(x_i)$ for each $i \in \omega$. Then β_i is an increasing sequence and must have a limit $\beta \leq \Phi(z)$. Since $\Phi[P]$ is closed, $\beta \in \Phi[P]$. It follows that $\beta = \Phi[y]$ and therefore $y \in P$. $\qquad\square$

Theorem 5.4.8. *Any well-ordered subset of $\mathbb{R}$ is countable.*

Proof. Let P be a subset of $\mathbb{R}$ well-ordered under the standard ordering; without loss of generality, we may assume that $P \subseteq (0, 1)$

and that P has no maximal element. Assume by way of contradiction that P is uncountable. For each $x \in P$, let x' be the successor of x in P, that is, the least element of $\{y \in P : x < y\}$. Then for each $x \in P$, $x' - x > 0$ and hence $x' - x > \frac{1}{n}$ for some positive integer n. Let $P_n = \{x \in P : x' - x > \frac{1}{n}\}$. Then $P = \bigcup_n P_n$. Since P is uncountable, it follows that, for some n, P_n is infinite. Now, let x_0 be the least element of P_n and, for each i, let x_{i+1} be the least element of P_n which is greater than x_i. Then, for each i, $x_{i+1} \geq x'_i$ and it follows that $x_{i+1} - x_i > \frac{1}{n}$. But this implies that $x_n - x_0 > 1$, contradicting our assumption that $P \subseteq (0,1)$. $\qquad\square$

Exercise for Section 5.4

Exercise 5.4.1. Suppose that $(A_0, R_0), (A_1, R_1), \ldots$ is an infinite sequence of well-ordered sets and each (A_i, R_i) is isomorphic to an ordinal α_i. Then $\bigoplus_i (A_i, R_i)$ is isomorphic to $\sum_{i < \omega} \alpha_i$.

Exercise 5.4.2. Find a well-ordered set of rationals with order type $\omega^2 + \omega$.

Exercise 5.4.3. Find a well-ordered set of rationals with order type ω^3.

Chapter 6

Cardinality and the Axiom of Choice

In this chapter, we present several principles which are equivalent to the Axiom of Choice, as well as applications of these principles. In particular, the Axiom of Choice implies that every set is isomorphic to a cardinal number and this makes working with cardinality much smoother. We define cofinality and regularity, limit and successor cardinals. Then we introduce inaccessible cardinals, which are the beginning of a vast array of large cardinals, which play a very important role in set theory. There is also an introduction to cardinal arithmetic.

6.1 Equivalent Versions of the Axiom of Choice

In this section, we look at several equivalent versions of the Axiom of Choice. To prove that these are equivalent, we appeal to transfinite recursion. The first version is the famous Well-Ordering Principle of Zermelo [12].

Definition 6.1.1. The *Well-Ordering Principle* is the statement "every set can be well-ordered".

Theorem 6.1.2 (Zermelo). *The following are equivalent on the basis of ZF axioms*:

1. *the Axiom of Choice*;
2. *the Well-Ordering Principle*.

Proof. (1) implies (2) is the more difficult implication. Assume the Axiom of Choice. Let x be an arbitrary set. It is enough to show that there is a bijection between x and an ordinal. Let h be a selector function on $\mathcal{P}(x) \setminus \{0\}$ as guaranteed by the Axiom of Choice. By transfinite recursion define a class function G on ordinals by $G(\alpha) = h(x \setminus \mathrm{rng}(G \upharpoonright \alpha))$ if the set $x \setminus \mathrm{rng}(G \upharpoonright \alpha)$ is non-empty, and $G(\alpha) = \texttt{trash}$ otherwise.

There must be an ordinal β such that $G(\beta) = \texttt{trash}$, otherwise G would be an injection from the proper class of all ordinals to the set x. Such injections do not exist though by the result of Exercise 3.7.3. Let β be the smallest ordinal such that $G \upharpoonright \beta = \texttt{trash}$. We will show that $G \upharpoonright \beta$ is a bijection between x and β. This will prove (2).

First of all, $G \upharpoonright \beta$ is a function with domain β by its definition. Its range must be equal to x, since there is no other way that $G(\beta) = \texttt{trash}$ could occur. Finally, $G \upharpoonright \beta$ is an injection. If this failed, there would have to be ordinals $\delta \in \gamma \in \beta$ such that $G(\delta) = G(\gamma) \in x$; however, this contradicts the recursive definition of the value $G(\gamma)$ which cannot belong to $\mathrm{rng}(G \upharpoonright \gamma)$, and therefore cannot be equal to $G(\delta)$.

To prove that (2) implies (1), assume that the Well-Ordering Principle holds. To verify the Axiom of Choice, let x be a collection of nonempty sets. To produce a selector on x, just use the Well-Ordering Principle to find a well-ordering on $\bigcup x$, and let f be the function such that $\mathrm{Dmn}(f) = x$ and $f(y)$ is the $\le$-least element of y, whenever $y \in x$. This proves (1). □

Now, we come to another equivalent of the Axiom of Choice, the very important *Zorn's Lemma*. It is the most commonly used form of the Axiom of Choice in mathematics, since its use does not require technical tools such as transfinite recursion.

Recall that an element p of a partially ordered set $(P, \le)$ is maximal if there is no element $q \in P$ strictly larger than p. An element p of P is an upper bound for a subset A of P if $q \le p$ for every $q \in A$.

Definition 6.1.3. *Zorn's Lemma* is the following statement. Whenever $(P, \le)$ is a non-empty partially ordered set such that every linearly ordered subset of P has an upper bound, then P has a maximal element.

Theorem 6.1.4 (Kuratowski [5]). *The following are equivalent on the basis of axioms of ZF set theory:*

1. *the Axiom of Choice;*
2. *Zorn's Lemma.*

Proof. We start with the implication (1) → (2). Let P be a partially ordered set. Let **trash** be a set which is not an element of P. Use the Axiom of Choice to find a selector h on the set $\mathcal{P}(P) \setminus \{0\}$. By transfinite recursion define a class function G on ordinals by the equation $G(\alpha) = h(a_\alpha)$, where $a_\alpha = \{p \in P : (\forall \beta \in \alpha)\, G(\beta) < p\}$ if the set $a_\alpha \subseteq P$ is non-empty, and $G(\alpha) = $ **trash** otherwise.

As in the previous proofs, there must be an ordinal β such that $G(\beta) = $ **trash**; otherwise, the function G would be an injection from the proper class of all ordinals to the set P, an impossibility by Exercise 3.7.3. Let β be the $\in$-smallest ordinal such that $G(\beta) = $ **trash**. We prove that β is a successor ordinal, $\beta = \gamma + 1$ for some γ, and $G(\gamma)$ is a maximal element of P.

To this end, observe that the recursion formula implies that the map $G \restriction \beta$ is a strictly increasing function from β to P; every value of G is larger than all the previous values. As a result, the set $G[\beta] \subseteq P$ is linearly ordered, and by the assumption on the partial ordering P, it has an upper bound p. Note that necessarily $p \in G[\beta]$ must hold because otherwise the set a_β is nonempty, containing at least p, and then $G(\beta)$ would not be equal to **trash**. The only way that $p \in G[\beta]$ can occur is that there is a largest ordinal $\gamma \in \beta$, and $G(\gamma) = p$.

To show that p is maximal in P, suppose for contradiction that it is not and that there is a strictly larger element $r \in P$. Then, the set a_β is non-empty, containing at least r, and so $G(\beta)$ would not be equal to **trash**. This contradiction completes the proof of that (1) implies (2).

For the implication (2)→(1), assume that Zorn's Lemma holds. Let x be a set of non-empty sets. To confirm the Axiom of Choice, we must produce a selector for x. Consider the partially ordered set P of all functions f such that $Dmn(f) \subseteq x$, and for all $y \in Dmn(f)$, $f(y) \in y$. The ordering on P is inclusion: $f \leq g$ if $f \subseteq g$. Every linearly ordered subset of P has an upper bound: If $a \subseteq P$ is a collection linearly ordered by inclusion, then $\bigcup a \in P$ is the upper bound. By an application of Zorn's lemma, the partially ordered set

P must have a maximal element, call it h. We show that h is a selector on x.

Indeed, suppose for contradiction that h is not a selector on x. The only way that can happen is that $Dmn(h) \neq x$. Let $y \in x$ be some set not in the domain of h. Let $z \in y$ be an arbitrary element. Consider the set $f = h \cup \{(y, z)\}$. It is clear that f is an element of the partially ordered set P, $h \subseteq f$, and $h \neq f$. This contradicts the maximal choice of h and completes the proof of the theorem. □

There are two variations of Zorn's Lemma which turn out to be equivalent to the Axiom of Choice.

Definition 6.1.5. *Hausdorff's Maximal Principle* states that any chain in a partially ordered set may be extended to a maximal chain.

Definition 6.1.6. *Kuratowski's Principle* is the following statement: Let Z be a family of sets, such that for every linearly ordered family $C \subseteq Z$, $\bigcup C \in Z$. Then Z contains a maximal set, under inclusion.

Kuratowski's principle follows from Zorn's Lemma as a special case when the partial order is given by set inclusion.

Exercises for Section 6.1

Exercise 6.1.1. Show that Kuratowski's Principle is equivalent to Zorn's Lemma. *Hint*: Examine the proof of Theorem 6.1.4.

Exercise 6.1.2. Show that Hausdorff's Maximal Principle is equivalent to Zorn's Lemma. *Hint*: Consider the family of chains partially ordered by set inclusion.

Exercise 6.1.3. Let A be a p.o. set such that every chain in A has a maximal element and let $F : A \to A$ be a function such that $f(a) \geq a$ for all $a \in A$. Prove that F has a *fixed point* c such that $f(c) = c$.

6.2 Applications of the Axiom of Choice

Since Zorn's Lemma is such a common presence in many mathematical arguments, at least one application of it is called for. Note the typical form of the argument: A complicated object is to be constructed.

The partially ordered set to which Zorn's Lemma is applied consists of approximations to such an object, and a maximal approximation (granted by Zorn's Lemma) is the object that we want. We provide some typical examples of this procedure.

Recall that a *basis* for a vector space V over a field F may be characterized as a maximal independent set. Here we say that a set B of vectors is independent if there is no non-trivial finite linear combination $c_1 v_1 + \cdots + c_n v_n$ of vectors from B, with coefficients from F, which equals the zero vector. We say that a set B of vectors from V is a spanning set if any $v \in V$ may be represented as a linear combination $c_1 v_1 + \cdots + c_n v_n$ of vectors from B. The usual definition of a basis is a set which is both independent and spanning. It is easy to see that a maximal independent set must also be a spanning set and hence a basis.

Theorem 6.2.1. *(AC) Every vector space V has a basis.*

Proof. Let $(P, \subseteq)$ be the partially ordered set of independent sets of vectors from V, ordered by inclusion. Let us verify that the union of a chain of independent sets is independent. Suppose that C is a chain of independent sets of vectors and let $v_1, \ldots, v_n \in \bigcup C$. For each i, there are $A_i \in C$ with $v_i \in A_i$. Since C is a chain, there must be some k such that each $A_i \subseteq A_k$, and hence each $v_i \in A_k$. Since A_k is independent, there can be no non-trivial linear combination $c_1 v_1 + \cdots + c_n v_n = 0$. Thus, $\bigcup C$ is independent. Now, by Zorn's Lemma, P has a maximal element, that is, a maximal independent set. $\quad\square$

Definition 6.2.2. Let x be a set. A *filter* on x is a set $F \subseteq \mathcal{P}(x)$ which is closed under supersets, that is,

$$(\forall y \in F)(\forall z \subseteq x)[y \subseteq z \to z \in F],$$

and closed under intersections, that is,

$$(\forall y \in F)(\forall z \in F)[y \cap z \in F],$$

and does not contain the empty set. An *ideal* on x is a set $I \subseteq \mathcal{P}(x)$ which is closed under subsets and unions and does not contain x.

It should be clear that the notions of filter and ideal are in a sense dual: If F is a filter on a set x, then $I = \{x \setminus y : y \in F\}$ is an ideal on x, and moreover, if I is an ideal on a set x, then $F = \{x \setminus y : y \in I\}$

is a filter on x. A filter typically serves as a measure of largeness of a subset of x, while an ideal serves as a notion of smallness.

Example 6.2.3. The *Fréchet ideal* on an infinite set x is the collection of all finite subsets of x.

Example 6.2.4. The *density zero ideal* on ω is the set of all sets $a \subseteq \omega$ whose upper asymptotic density $\limsup_n \frac{|a \cap n|}{n}$ is equal to zero.

In many circumstances, one would like to use a filter on a set x which for every set $y \subseteq x$ decides whether y is large or small, as in the following definition:

Definition 6.2.5. A filter F on a set x is an *ultrafilter* if for every set $y \subseteq x$, $y \in F$ or $x \setminus y \in F$. The ideal dual to an ultrafilter is a *maximal ideal*. An ideal is said to be *prime* if whenever $a \vee b \in I$, then either $a \in I$ or $b \in I$.

Note that an ideal I on a set x is prime if and only if it is maximal. This is Exercise 6.2.4. The catch is, how do we find an ultrafilter? There is a rather obvious and useless type of ultrafilter: the *principal* kind. An ultrafilter F is principal if there is an element $i \in x$ such that $y \in F$ if and only if $i \in y$. Are there any non-principal ultrafilters? This is equivalent to the existence of a maximal ideal. The Axiom of Choice yields a positive answer:

Theorem 6.2.6 (Prime Ideal Theorem). *(AC) There is a non-principal ultrafilter on every infinite set.*

Proof. Let x be an infinite set. Let P be the poset of all filters on x which do not contain any finite sets. The ordering on P is inclusion. We use Zorn's Lemma to produce a maximal element in P. Then, we show that this maximal element is a non-principal ultrafilter.

First, observe that P is a non-empty poset. For this, consider $F = \{y \subseteq x : x \setminus y \text{ is finite}\}$. It is easy to check that F is a filter. Since x is infinite, $0 \notin F$. Since the union of finite sets is finite, F is closed under intersections. As a subset of a finite set is finite again, F is closed under supersets. Lastly, since x is infinite, F contains no finite sets.

Second, observe that every linearly ordered set $a \subseteq P$ has an upper bound. This upper bound is $\bigcup a$. To verify that $\bigcup a$ is indeed an element of P, observe the following:

- $\bigcup a$ contains no finite sets as no filters in a contain any finite sets.
- To check the closure of a under supersets, let $y \subseteq x$ be an element of $\bigcup a$ and $y \subseteq z$ be a subset of x. Choose $F \in a$ such that $y \in F$. Since F is a filter, $z \in F$ and so $z \in \bigcup a$.
- To check the closure of $\bigcup a$ under intersections, we finally use linearity of a. Suppose that $y, z \in \bigcup a$ and $F, G \in a$ are such that $y \in F$ and $z \in G$. By linearity of a, either $F \subseteq G$ or $G \subseteq F$ holds. For definiteness, suppose $F \subseteq G$. Then $y \in G$, and since G is a filter closed under intersections, $y \cap z \in G$ and so $y \cap z \in \bigcup a$ as required.

Now, Zorn's Lemma shows that the poset P has a maximal element F. Let $x = y \cup z$ be a partition; we show that either $y \in F$ or $z \in F$.

Claim 6.2.7. *Either $\forall u \in F$ $u \cap y$ is infinite or $\forall u \in F$ $u \cap z$ is infinite.*

Proof. If both of the disjuncts failed, then there would be sets $u_y, u_z \in F$ such that $u_y \cap y$ is finite and $u_z \cap z$ is finite. Consider the set $u = u_y \cap u_z$. Since F is closed under intersections, $u \in F$. Since $x = y \cup z$, it must be the case that $u \subseteq (u_y \cap y) \cup (u_z \cap z)$. This is a union of two finite sets and therefore finite. This contradicts the assumption that elements of P contain no finite sets. $\square$

Now, one of the disjuncts in the claim must hold; for definiteness, assume that, for all $u \in F$, $u \cap y$ is infinite. Consider $G = \{v \subseteq x \colon (\exists u \in F)\, u \cap y \subseteq v\}$. This is a filter containing no finite sets, containing F as a subset, and y as an element. By the maximality assumption, it must be the case that $F = G$. Thus, $y \in F$ as required. $\square$

Exercises for Section 6.2

Exercise 6.2.1. Let $(A, \leq)$ be a well-ordering on a set A. Let $B \subseteq A$ be any set. Prove that B equipped by the ordering inherited from A is again a well-ordering.

Exercise 6.2.2. Let $\leq$ be a linear ordering. Show that the following are equivalent:

1. $\leq$ is a well-ordering.
2. There is no infinite strictly descending sequence $x_0 > x_1 > x_2 > \cdots$ in the ordering given by $\leq$.

Exercise 6.2.3. Show that if an ultrafilter contains a finite set, then it is in fact principal.

Exercise 6.2.4.

1. Show that a filter F is an ultrafilter if and only if, for any two sets A and B, if $A \cup B \in F$, then either $A \in F$ or $B \in F$.
2. Show the dual result, that an ideal I is maximal if and only if for any two sets A and B, if $A \cap B \in I$, then either $A \in I$ or $B \in I$.

Exercise 6.2.5. For any non-empty subset C of A, $F_C = \{B \in \mathcal{P}(A) : C \subseteq B\}$ is a filter.

Exercise 6.2.6. Show that for a finite set A, every filter on F has the form F_C for some C.

Exercise 6.2.7. Show that every proper filter on a set x can be extended to an ultrafilter.

Exercise 6.2.8. Let $(P, \leq)$ be a partial ordering. Show that there is a set $A \subseteq P$ such that any two elements of A are incomparable in $\leq$ and for every $p \in P$ there is $q \in A$ such that p, q are comparable.

Exercise 6.2.9. Show that any partial ordering on a set A may be extended to a linear ordering of A.

Exercise 6.2.10. Show that any ideal I in a ring R with unity may be extended to a maximal ideal, that is, an ideal M such that there is no ideal J with $M \subseteq J \subseteq R$ except for M and R.

6.3 Cardinal Numbers

The purpose of this section is to further develop the theory of cardinalities under the Axiom of Choice. In particular, we identify a canonical representative for each cardinality and show that cardinalities are linearly ordered.

Definition 6.3.1. A *cardinal number*, or cardinal for short, is an ordinal number which is not in a bijective correspondence with any ordinal number smaller than it.

In particular, every natural number as well as ω is a cardinal number. In set-theoretic literature, cardinals are typically denoted by lowercase Greek letters such as $\kappa, \lambda, \mu, \ldots$.

Example 6.3.2. No infinite successor ordinal can be a cardinal. To see this, let $\alpha \geq \omega$ be an ordinal. Then we may define a bijection from α onto $\alpha + 1 = \alpha \cup \{\alpha\}$ by mapping 0 to α, mapping $n + 1$ to n for $n \in \omega$, and mapping β to β for $\omega \leq \beta < \alpha$.

Recall that in Definition 4.4.1 of Section 4.4, the notion of cardinality was defined as an equivalence relation. We defined

$$|x| = |y| \iff \text{there is an bijection from } x \text{ to } y.$$

Thus, we may view $|x|$ as the equivalence class of x under the cardinality equivalence relation. The following theorem shows that, assuming the Axiom of Choice, every set is equivalent to a cardinal number under this equivalence relation.

Theorem 6.3.3. *(AC) Every set is a bijective image of a unique cardinal number.*

Proof. Let x be any set. Let a be the class of all ordinal numbers which are bijective images of x. Observe that a is non-empty: By Zermelo's Well-Ordering Principle, x can be well-ordered and the well-ordering on it is isomorphic to some ordinal. The isomorphism is then a bijective function between x and the ordinal.

Now, the class a must have an $\in$-least element. By the definition of a, this minimum of a is a cardinal number. This shows that x is in bijective correspondence with some cardinal number. The uniqueness of this cardinal number follows easily: If κ, λ are cardinals such that $|\kappa| = |x| = |\lambda|$, then κ and λ are in a bijective correspondence. This excludes both $\kappa \in \lambda$ and $\lambda \in \kappa$ by the definition of a cardinal number, and by the linearity of ordering of the ordinal numbers (Theorem 5.1.3), $\kappa = \lambda$ is the only option left. $\square$

We define the cardinality of a set x to be the unique cardinal given by Theorem 6.3.3.

Lemma 6.3.4.

1. *For any cardinals κ and λ, $|\kappa| \leq |\lambda| \iff \kappa \leq \lambda$.*
2. *For any sets x, y and any cardinals κ, λ such that $|x| = \kappa$ and $|y| = \lambda$, $|x| \leq |y|$ if and only if $\kappa \leq \lambda$.*

Proof. Let sets x, y and cardinals κ and λ be given.

For part (1), suppose first that $|\kappa| \leq |\lambda|$. Then there is an injection $h : \kappa \to \lambda$. Let $A = h[\kappa] \subseteq \lambda$ and let α be the order type of $(A, \in)$. Now, κ is set isomorphic to A via the map h and hence is set isomorphic to α. Since κ is a cardinal, it follows that $\kappa \leq \alpha$. Now, $\alpha \leq \lambda$ by Proposition 5.4.2, and therefore $\kappa \leq \lambda$.

Suppose next that $\kappa \leq \lambda$. Then the identity is an injection from κ into λ and hence $|\kappa| \leq |\lambda|$.

For part (2), let $f : \kappa \to x$ and $g : y \to \lambda$ be set isomorphisms. For the first direction, suppose that $|x| \leq |y|$. Then there is an injection $i : x \to y$. It follows that $g \circ i \circ f : \kappa \to \lambda$ is an injection, and therefore $|\kappa| \leq |\lambda|$. Then by part (1), $\kappa \leq \lambda$.

For the other direction, suppose that $\kappa \leq \lambda$. Then $g^{-1} \circ f^{-1} : x \to y$ is an injection so that $|x| \leq |y|$. $\qquad \square$

Recall the Trichotomy Property of a linear order $\leq$ with corresponding strict order $<$ states that, for any x, y, exactly one of the following holds: $x < y$, $y < x$, or $x = y$. For the relation $|x| \leq |y|$ on the cardinality classes, the relation $|x| < |y|$ is given by $|x| \leq |y|$ but not $|y| \leq |x|$, that is, there is an injection from x into y but no injection from y into x. Here is the Trichotomy Property for cardinality.

Corollary 6.3.5. *(AC) Whenever x, y are sets, then exactly one of the following holds:*

1. $|x| < |y|$;
2. $|y| < |x|$;
3. $|x| = |y|$.

Proof. Let sets x, y be given. Let κ, λ be cardinals such that $|\kappa| = |x|$ and $|\lambda| = |y|$. Since the ordinal numbers are linearly ordered, by Theorem 5.1.3, it follows that exactly one of the following holds:

$\kappa < \lambda$, $\lambda < \kappa$, or $\lambda = \kappa$. Now, by Lemma 6.3.4, we have $|x| < |y| \iff$ $\kappa < \lambda$, $|y| < |x| \iff \lambda < \kappa$, and $|x| = |y| \iff \kappa = \lambda$. $\quad\square$

Thus, under the Axiom of Choice, cardinalities are linearly ordered (even well-ordered), and the cardinal numbers are canonical representatives of cardinalities.

We now pause to consider the converse of Corollary 6.3.5. Here are two more statements equivalent to the Axiom of Choice:

- *The Injection Principle*: For any sets A and B, there is either an injection from A into B or there is an injection from B into A.
- *The Mapping Principle*: For any sets A and B, there is either a surjection from A into B or there is a surjection from B onto A.

It is easy to see that whenever there is an injection from A into B, then there is a surjection from B onto A. This is Exercise 6.3.2. It follows that the Injection Principle implies the Mapping Principle.

The following lemmas are connected with showing that the Mapping Principle and the Injection Principle are equivalent to the Axiom of Choice.

Lemma 6.3.6. *For any set x, let $\Gamma(x)$ be the set of ordinals α such that $|\alpha| \leq x\}$. Then $\Gamma(x)$ is a set and is in fact the least ordinal which does not map 1-to-1 into x.*

Proof. By definition, $|\alpha| \leq x$ if and only if there is an injection from α to x, which will induce a well-ordering on a subset of x. Thus, the set of well-orderings on subsets of x maps onto $\Gamma(x)$, making it a set by Replacement. The second conclusion is left as an exercise. $\quad\square$

We observe that $\Gamma(\omega)$ will be the first uncountable cardinal $\aleph_1$. A more general result is given in 6.3.10.

Lemma 6.3.7. *If there is a mapping from A onto B, then there is an injection from $P(B)$ to $P(A)$.*

Proof. Suppose g maps A onto B. Definite the mapping $f : \mathcal{P}(B) \to \mathcal{P}(A)$ by

$$f(t) = g^{-1}(t) = \{a \in A : g(a) \in t\}.$$

To see that this is an injection, suppose that $s \neq t$. Without loss of generality, there is some $b \in s \setminus t$; let $g(a) = b$. Then $a \in f(s)$, but $a \notin f(t)$. So, $f(s) \neq f(t)$. □

Proposition 6.3.8. *The following are equivalent*:

1. *The Axiom of Choice*;
2. *The Injection Principle*;
3. *The Mapping Principle*.

Proof. We have (1) implies (2) by Corollary 6.3.5 and we have (2) implies (3) by Exercise 6.3.2. It remains to show that (3) implies (1). Assume the Mapping Principle and let A be an arbitrary set. We show that A can be well ordered. This implies the Axiom of Choice, by Theorem 6.1.2. It follows from the Mapping Principle that, for any ordinal α, there is a surjection g_α either from α onto A or from A onto α. Suppose first that there is some α such that g_α maps α onto A. Then define an injection from A into α by letting $f(x)$ be the least $\beta < \alpha$ such that $g(\beta) = x$. This induces a well-ordering of A, where $a < b \iff f(a) < f(b)$.

Next, suppose that, for every ordinal α, g_α maps A onto α. For each α, let $A_\alpha = \{g_\alpha^{-1}(\beta) : \beta < \alpha\}$ be well ordered by having $g_\alpha^{-1}(\beta) < g_\alpha^{-1}(\gamma) \iff \beta < \gamma$. Each A_α is a subset of $\mathcal{P}(A)$ and has order type α, so for each α, there is a well-ordering of a subset of $\mathcal{P}(A)$ of order type α. Now, let W_A be the set of well-orderings of subsets of $\mathcal{P}(A)$. Then the mapping taking each $R \in W_A$ to its order type maps the set W_A onto the class of ordinals, violating the Axiom of Replacement. □

In order to justify the existence of the first infinite ordinal, ω, as a set, it was necessary to introduce the Axiom of Infinity. We have now obtained some understanding of the countable ordinals, such as ω, $\omega \cdot 3$, ω^5, and so on. Thus, we can imagine the first uncountable cardinal ω_1 (also known as $\aleph_1$) as the class of all countable ordinals. The question now is whether another axiom is needed to justify that ω_1 exists as a set. It turns out that the Axiom of Replacement is sufficient to prove this. There is in fact an enormous supply of cardinal numbers, as described in the following theorem.

Lemma 6.3.9. *For any cardinal number κ and any ordinal λ, λ has cardinality κ if and only if there is a well-ordering of κ which has order type λ.*

Proof. Suppose first that R is a well-ordering of κ of order type λ. Then the function $G : \lambda \to \kappa$ described in Theorem 5.4.1 is an isomorphism, showing that κ and λ have the same cardinality κ.

Suppose next that λ has cardinality κ. Then there must be an isomorphism $G : \kappa \to \lambda$. This will induce a well-ordering R on κ of type λ, where $\alpha R \beta \iff G(\alpha) < G(\beta)$. $\qquad\square$

Theorem 6.3.10 (Hartog's Lemma). *For any cardinal number κ, there exists a cardinal κ^+ which is the immediate successor of κ, as a cardinal. That is, $\kappa < \kappa^+$ and there is no cardinal λ with $\kappa < \lambda < \kappa^+$.*

Proof. If κ is finite, then $\kappa^+ = \kappa + 1$. Let the cardinal number $\kappa \geq \omega$ be given and let A be the set of well-orderings R of κ. Since κ is a limit ordinal, each pair in R belongs to V_κ so that $R \subseteq V_\kappa$ and hence $R \in V_{\kappa+1}$. So, $A \in V_{\kappa+2}$ is indeed a set by Comprehension. Now, consider the function G mapping each well-ordering R in A to its order type. The image $G[A]$ must be the set of all ordinals having cardinality κ. We claim that $\gamma = \bigcup G[A]$ is the desired cardinal κ^+. Certainly, $\kappa < \gamma$ since the standard ordering on κ has order type κ. If $\kappa < \lambda < \gamma$, then by definition of γ, there must be a well-ordering of κ of type λ, and hence λ has cardinality κ by Lemma 6.3.9. $\qquad\square$

It follows from Theorem 6.3.10 that the cardinal ω_1 equals ω^+ and hence is a set.

There is an alternative proof of the existence of uncountable cardinals using the Axiom of Choice. Fix a cardinal λ. By Theorem 4.4.8, $|\mathcal{P}(\lambda)| > \lambda$. By the Axiom of Choice, there is a cardinal number κ such that $|\kappa| = |\mathcal{P}(\lambda)| = 2^\lambda$ and therefore $\lambda < \kappa$.

We may now use Theorem 5.3.1 to define an enumeration of the cardinals as follows:

Definition 6.3.11. The class function mapping ordinals to cardinals is given by the following:

1. $\aleph_0 = \omega$.
2. For each ordinal α, $\aleph_{\alpha+1} = \aleph_\alpha^+$.
3. For each limit ordinal λ, $\aleph_\lambda = \bigcup_{\alpha<\lambda} \aleph_\alpha$.

It follows by induction on ordinals that every infinite cardinal equals $\aleph_\beta$ for some ordinal β.

An alternative sequence of cardinals may be defined using 2^κ in place of κ^+, as follows:

Definition 6.3.12. The class function $\beth_\alpha$ mapping ordinals to cardinals is given by the following:

1. $\beth_0 = \omega$.
2. For each ordinal α, $\beth_{\alpha+1} = 2^{\beth_\alpha}$.
3. For each limit ordinal λ, $\beth_\lambda = \bigcup_{\alpha<\lambda} \beth_\alpha$.

Definition 6.3.13. Let λ be an infinite cardinal.

1. λ is *successor cardinal* if $\lambda = \kappa^+$ for some cardinal κ.
2. λ is a *limit cardinal* if for all cardinals $\kappa < \lambda$, $\kappa^+ < \lambda$.
3. λ is a *strong limit cardinal* if, for all cardinals $\kappa < \lambda$, $2^\kappa < \lambda$.

Example 6.3.14. For each n, $\aleph_{n+1} = \aleph_n^+$ is a successor cardinal. The cardinal $\aleph_\omega = \bigcup_n \aleph_n$ is a limit cardinal. The cardinal $\beth_\omega = \bigcup_n \beth_n$ is a strong limit cardinal.

Next, we briefly consider some cardinal arithmetic.

Definition 6.3.15. For two cardinals κ and λ, the cardinal sum $\kappa + \lambda = |\kappa \oplus \lambda|$ and the the cardinal product $\kappa \cdot \lambda = |\kappa \times \lambda|$. The cardinal exponent κ^λ is the cardinality of the set κ^λ of functions mapping from λ into κ.

As we saw in Propositions 5.3.5 and 5.3.9, this means that the *cardinal* sum and product of two cardinals are equal to the cardinality of the *ordinal* sum and product. However, the ordinal exponent $2^\omega = \omega$, whereas the cardinal exponent 2^ω is uncountable.

Proposition 6.3.16. *For any infinite cardinal κ, the cardinal sum $\kappa + \kappa = \kappa$.*

Proof. For the first equality, let κ be any limit ordinal and define the bijection $f : \kappa \oplus \kappa \to \kappa$ by $f(0, x) = 2 \cdot x$ and $f(1, x) = 2 \cdot x + 1$. We leave it as an exercise to show that this is one-to-one and onto. $\square$

Proposition 6.3.17. *For any infinite cardinal κ, the cardinal product $\kappa \cdot \kappa = \kappa$.*

Proof. The proof is by transfinite induction. For $\kappa = \aleph_0 = \omega$, we have an injection from $\omega \times \omega$ into ω given by $f(x, y) = 2^x \cdot 3^y$.

For $\kappa > \aleph_0$, suppose by induction that $\lambda \cdot \lambda = \lambda$ for all cardinals $\lambda < \kappa$. Define a well-ordering on $\kappa \times \kappa$ by setting $(\alpha_1, \beta_1) < (\alpha_2, \beta_2)$ if and only if

(1) $\max\{\alpha_1, \beta_1\} < \max\{\alpha_2, \beta_2\}$, or
(2) $\max\{\alpha_1, \beta_1\} = \max\{\alpha_2, \beta_2\}$ and $\alpha_1 < \alpha_2$, or
(3) $\max\{\alpha_1, \beta_1\} = \max\{\alpha_2, \beta_2\}$ and $\alpha_1 = \alpha_2$ and $\beta_1 < \beta_2$.

Note that when $max\{\alpha_1, \beta_1\} = max\{\alpha_2, \beta_2\}$, the other two cases just say that (α_1, β_1) precedes (α_2, β_2) in the lexicographic order.

We claim that the order type of this well-ordering is exactly κ so that the natural mapping from $\kappa \times \kappa$ to the ordinals defined by this well-ordering is an injection from $\kappa \times \kappa$ into κ. This shows that $\kappa \cdot \kappa = \kappa$, as desired.

To check this claim, note that for each fixed pair (α, β), there are at most $|\max\{\alpha, \beta\} + 1)|^2$ predecessors of (α, β), and $|\max\{\alpha, \beta\} + 1)|^2 < \kappa$ by induction. Thus, each initial segment of the well-order has order type $< \kappa$. $\square$

In Exercise 6.3.12, we consider an alternative proof of Proposition 6.3.17 using ordinal arithmetic.

More generally, we have the following:

Proposition 6.3.18. *For any infinite cardinals κ and λ, $\kappa + \lambda = \kappa \cdot \lambda = \max\{\kappa, \lambda\}$.*

We can now generalize the result from Theorem 4.5.2 that (assuming AC) a countable union of countable sets is countable.

Theorem 6.3.19. *(AC) For any cardinals κ and λ and any sets $\{A_\alpha : \alpha < \lambda\}$ such that $|A_\alpha| \leq \kappa$ for all $\alpha < \lambda$, $|\bigcup_{\alpha < \lambda}| \leq \kappa \cdot \lambda$.*

Proof. Let κ, λ, and $\{A_\alpha : \alpha < \lambda\}$ be given as above and let $A = \bigcup_{\alpha < \lambda} A_\alpha$. For each $\alpha < \lambda$, let F_α be an injection from A_α into κ. For each $a \in A$, let $\alpha(a)$ be the least $\alpha < \lambda$ such that $a \in A_\alpha$ and let $G(a) = (\alpha(a), F_{\alpha(a)}(a))$. Then G is an injection from A into $\lambda \times \kappa$. It follows that $|A| \leq \lambda \cdot \kappa$. $\square$

In particular, when $\kappa = \lambda$, we see that the union of κ sets, each of cardinality $\leq \kappa$, has cardinality $\leq \kappa$.

We can also define infinite sums and products of cardinal numbers, as we did for ordinal numbers.

Definition 6.3.20. For any indexed family $\{\kappa_i : i \in I\}$ of cardinals, the sum $\sum_{i \in I} \kappa_i = |\bigoplus_{i \in I} \kappa_i|$ and $\prod_{i \in I} \kappa_i = |\prod_{i \in I} \kappa_i|$.

Example 6.3.21. $\sum_{i < \omega} \aleph_i = \aleph_\omega$. To see this, we need to establish the inequality in both directions. Certainly, $\aleph_j \leq \sum_{i < \omega} \aleph_i$ for each $j < \omega$ and it follows that $\aleph_\omega \leq \sum_{i < \omega} \aleph_i$. For the other direction, we can define an embedding of $\bigoplus_{i < \omega} \aleph_i$ into $\aleph_\omega$ by mapping (i, α) to $\aleph_i + \alpha$.

Example 6.3.22. $1 \cdot 2 \cdot 3 \cdots = \prod_{i < \omega} i + 1 = 2^{\aleph_0}$. For one inequality, define an embedding from $\{f : \omega \to \{0,1\}\}$ into $\prod_{i < \omega} i + 1$ by mapping f to the function g, where $g(0) = 0$, and $g(i + 1) = f(i)$. Thus, $\prod_{i < \omega} i + 1 \geq 2^{\aleph_0}$. The other inequality is left as an exercise.

Theorem 6.3.23. *For any indexed families $\{\kappa_i : i \in I\}$ and $\{\lambda_i : i \in I\}$ of cardinals, if $\kappa_i < \lambda_i$ for all $i \in I$, then $\sum_{i \in I} \kappa_i < \prod_{i \in I} \lambda_i$.*

Proof. This is essentially a diagonal argument. Let $F : \bigoplus_{i \in I} \kappa_i \to \prod_{i \in I} \lambda_i$ be a possible isomorphism. For each i and $\alpha < \kappa_i$, let $F_{i,\alpha}$ denote $F(i, \alpha)$. Consider the mapping which takes $\alpha < \kappa_i$ to $F_{i,\alpha}(i) \in \lambda_i$. Since $\kappa_i < \lambda_i$, $F_{i,\alpha}$ cannot be onto, so for each i, there is some $\beta_i < \lambda_i$ such that $F_{i,\alpha}(i) \neq \beta_i$ for any $\alpha < \kappa_i$. Now, define $f \in \prod_{i \in I} \lambda_i$ by $f(i) = \beta_i$. We claim that f is not in the range of F. Suppose by way of contradiction that $f = F_{i,\alpha}$ for some i and $\alpha < \kappa_i$. Then we would have $F_{i,\alpha}(i) = \beta_i$, contradicting the choice of β_i. $\qquad\square$

Finally, we come to the formulation of the question which was one of the driving forces behind the development of modern set theory from its beginnings. The original Continuum Hypothesis conjectured that $\aleph_1 = 2^\omega$, that is, $\beth_1 = \aleph_1$. The Generalized Continuum Hypothesis conjectured that $\beth_\alpha = \aleph_\alpha$ for all α, which is equivalent to saying that $2^\kappa = \kappa^+$ for all infinite cardinals κ.

Question 6.3.24.

(i) The continuum problem: *Determine the ordinal α such that $|\mathcal{P}(\omega)| = \aleph_\alpha$.*

(ii) The generalized continuum problem: *For every ordinal α, determine the ordinal β such that $|\mathcal{P}(\aleph_\alpha)| = \aleph_\beta$.*

It turns out that the continuum problem cannot be resolved in ZFC. There is a good amount of speculation, some primitive and some highly sophisticated, as to what the "right" answer to the continuum problem "should" be. We can rule out one possibility.

Corollary 6.3.25. $\aleph_\omega \neq 2^{\aleph_0}$.

Proof. Suppose that $\aleph_n < 2^{\aleph_0}$ for all $n < \omega$. Then by Theorem 6.3.23,

$$\sum_{n<\omega} \aleph_n < \prod_{n<\omega} 2^{\aleph_0}.$$

By Example 6.3.21, the left-hand side equals $\aleph_\omega$. The right-hand side equals $(2^{\aleph_0})^{\aleph_0} = 2^{\aleph_0 \cdot \aleph_0} = 2^{\aleph_0}$. $\qquad\square$

Before we leave the subject of cardinal numbers, we develop the notion of cofinality.

Definition 6.3.26. Let α and β be limit ordinals.

1. A subset x of β is *cofinal* in β if $\bigcup x = \beta$, that is, for every $\alpha < \beta$, there exists $y \in x$ such that $\alpha < y$.
2. The cofinality $\mathsf{cof}(\beta)$ is the least cardinal α such that there is a function $f : \alpha \to \beta$ with $Rng(f)$ cofinal in β.
3. The ordinal β is *regular* if $\mathsf{cof}(\beta) = \beta$.
4. β is *singular* if $\mathsf{cof}(\beta) < \beta$, that is, if β is not regular.

It is fairly immediate to observe that cofinality of any limit ordinal must be regular, and every regular ordinal is a cardinal. Certainly, $\aleph_0$ is a regular cardinal.

Example 6.3.27. If a limit ordinal β is not a cardinal, then it is bijective with a smaller ordinal and hence is singular. It is easy to see that $\aleph_\omega$ has cofinality ω and hence is not regular.

Here is an important property of regular cardinals which is a stronger version of Theorem 6.3.19.

Theorem 6.3.28. *(AC) For any regular cardinal κ, any $\lambda < \kappa$, and any set $\{A_\alpha : \alpha < \lambda\}$ such that $|A_\alpha| < \kappa$ for all $\alpha < \lambda$, $|\bigcup_{\alpha<\lambda} A_\alpha| < \kappa$.*

Proof. Let κ and $\{A_\alpha : \alpha < \lambda\}$ be given as above and let $A = \bigcup_{\alpha < \lambda} A_\alpha$. Define $F : \lambda \to \kappa$ by $F(\alpha) = |A_\alpha|$. Since κ is regular, $Rng(F)$ cannot be cofinal in κ. Hence, there is some cardinal $\mu < \kappa$ such that $|A_\alpha| \leq \mu$ for all $\alpha < \lambda$. It follows from Theorem 6.3.19 that $|\bigcup_{\alpha < \lambda} A_\alpha| \leq \mu \cdot \lambda = \max\{\mu, \lambda\} < \kappa$. $\qquad\square$

Many cardinals are regular, as becomes obvious from the following theorem.

Theorem 6.3.29. *Every successor cardinal is regular.*

Proof. This theorem requires the Axiom of Choice for its proof; without the Axiom of Choice, it may even happen that every limit ordinal has cofinality equal to ω. We just show that ω_1 is regular.

Suppose for contradiction that ω_1 is singular. Then, its cofinality must be equal to $\omega = \omega_0$ and there has to be a function $f : \omega \to \omega_1$ whose range is cofinal in ω_1. Then, $\omega_1 = \bigcup_n f(n)$ is a countable union of countable sets. Such unions are countable by Theorem 4.5.2(4), contradicting the definition of ω_1 as the first uncountable cardinal. $\qquad\square$

The theorem immediately suggests the following question:

Question 6.3.30. *Is there an uncountable regular limit cardinal?*

The question was considered by Hausdorff in 1908 and later greatly expanded by Tarski. The question cannot be resolved in ZFC.

Definition 6.3.31. Let κ be an infinite cardinal.

1. κ is *weakly inaccessible* if it is regular and is a limit cardinal.
2. κ is *strongly inaccessible* if it is regular and is a strong limit cardinal.

Inaccessible cardinals are the beginning of a hierarchy of *large cardinals* which is one of the main tools of modern set theory. The notion of a strongly inaccessible cardinal is important in Chapter 8. In particular, we need the following result.

Theorem 6.3.32. *If κ is a strongly inaccessible cardinal, then $|V_\alpha| < \kappa$ for all $\alpha < \kappa$.*

Proof. The proof is by induction on $\alpha < \kappa$.

Base Step. For $\alpha = 0$, $V_0 = \emptyset$ and $|V_0| = 0 < \kappa$.

Successor Step. Suppose that $|V_\alpha| < \kappa$. Now, $V_{\alpha+1} = \mathcal{P}(V_\alpha)$ so that $|V_{\alpha+1}| = 2^{|V_\alpha|}$. Since κ is a strong limit, it follows that $2^{|V_\alpha|} < \kappa$.

Limit Step. Suppose that λ is a limit ordinal and that $|V_\alpha| < \kappa$ for all $\alpha < \lambda$. Now, $V_\lambda = \bigcup_{\alpha<\lambda} V_\alpha$. Since κ is regular, it follows from Theorem 6.3.28 that $|V_\lambda| < \kappa$. □

Exercises for Section 6.3

Exercise 6.3.1. Show that, under the equivalence relation $|x| = |y|$, each equivalence class $|x|$ is a class, but, if $x \neq \emptyset$, then $|x|$ not a set.

Exercise 6.3.2. Show that if there is an injection from a set A into a set B, then there is a surjection from B to A.

Exercise 6.3.3. Show that for any set C of cardinals, $\bigcup C$ is a cardinal.

Exercise 6.3.4. Finish the proof of Lemma 6.3.6 by showing that $\Gamma(x)$ is in fact the least ordinal which does not map 1–1 into x.

Exercise 6.3.5. Verify that $\gamma = \bigcup G[A]$ defined in the proof of Theorem 6.3.10 is an ordinal number and also a cardinal number.

Exercise 6.3.6. Show that for any ordinals α and β, if $\alpha < \aleph_{\beta+1}$, then $|\alpha| \leq \aleph_\beta$.

Exercise 6.3.7. Show that every cardinal equals $\aleph_\beta$ for some ordinal β.

Exercise 6.3.8. Show that for any ordinal α, $\alpha \leq \aleph_\alpha$.

Exercise 6.3.9. Construct a cardinal κ so that $\aleph_\kappa = \kappa$.

Exercise 6.3.10. Prove that the $\aleph$ function mapping any ordinal α to $\aleph_\alpha$ is normal, that is, strictly increasing and continuous.

Exercise 6.3.11. Complete the proof that for any infinite cardinal κ, $\kappa + \kappa = \kappa$.

Exercise 6.3.12. Here is an alternative proof of Proposition 6.3.17 using ordinal arithmetic. For $x, y < \kappa$, let $F(x, y) = (x + y)^2 + x$. Show that F is an injection from $\kappa \times \kappa$ into κ.

To see this, suppose that $(x_1, y_1) \neq (x_2, y_2)$ and show that $(x_1 + y_1)^2 + x_1 \neq (x_2 + y_2)^2 + x_2$. Note that there are two cases, depending on whether $x_1 + y_1 = x_2 + y_2$.

Exercise 6.3.13. Show that for any infinite cardinals κ and λ, $\kappa + \lambda = \kappa \cdot \lambda = \max\{\kappa, \lambda\}$.

Exercise 6.3.14. Show that $\prod_{i<\omega} i + 1 \leq 2^{\aleph_0}$.

Exercise 6.3.15. Show that the cardinals $\aleph_0^{\aleph_0}$ and $2^{\aleph_0}$ are equal. More generally, show that for any cardinals κ and λ, if $2 \leq \kappa \leq 2^\lambda$, then $2^\lambda = \kappa^\lambda$.

Exercise 6.3.16. Show that for any limit ordinal α, $cof(\aleph_\alpha = cof(\alpha)$.

Exercise 6.3.17. Show that for any limit cardinal κ, $\kappa < cof(2^\kappa)$.

Exercise 6.3.18. Show that a cardinal κ is regular if and only if, for any cardinal $\lambda < \kappa$, and any set $\{A_\alpha : \alpha < \lambda\}$ such that $|A_\alpha| \leq \kappa$ for all $\alpha < \lambda$, $|\bigcup_{\alpha<\lambda} A_\alpha| < \kappa$.

Exercise 6.3.19. Show that $\aleph_0$ is a regular cardinal.

Exercise 6.3.20. Prove that every successor cardinal is regular.

Exercise 6.3.21. Verify that $\gamma = \bigcup G[A]$ defined in the proof of Theorem 6.3.10 is an ordinal number and also a cardinal number.

Chapter 7

Real Numbers

In this chapter, we describe the construction of the integers, the rational numbers, and the real numbers using only the tools of set theory.

7.1 Integers and Rational Numbers

In this section, we obtain representations of the integers and the rational numbers within set theory, using equivalence relations.

The set of natural numbers has the drawback that it is not closed under additive inverses and, in general, not closed under subtraction. To remedy this, let the pair (m, n) of natural numbers represent the integer $m - n$. Then -2 is represented by $(0, 2)$ and also by $\langle 1, 3 \rangle$. With this in mind, define the equivalence relation E_Z on $\mathbb{N} \times \mathbb{N}$ by setting

$$(m_1, n_1) E_Z (m_2, n_2) \iff n_1 + m_2 = m_1 + n_2.$$

Then $(m_1, n_1) E_Z (m_2, n_2)$ if and only if $n_1 - m_1 = n_2 - m_2$. It is left to the exercises to show that this is an equivalence relation.

Definition 7.1.1. The set $\mathbb{Z}$ of integers is the family of equivalence classes of pairs of natural numbers under the relation E_Z.

The natural number n is represented in $\mathbb{Z}$ by the class $[(n, 0)]$.

Addition of integers may be defined by setting $[(m_1, n_1)] + [(m_2, n_2)] = [(m_1 + m_2, n_1 + n_2)]$. The definition of multiplication is left the exercises.

Having obtained the integers through subtraction of natural numbers, we may now use division to define rational numbers. Here we restrict to pairs (i, j) of integers such that $j \neq 0$ and let

$$(i_1, j_1) E_Q (i_2, j_2) \iff i_1 \cdot j_2 = j_1 \cdot i_2.$$

Then the pair (i, j) is meant to represent the rational number i/j.

To define the standard linear ordering on $\mathbb{Z}$ and $\mathbb{Q}$, we first define the *positive* integers and rationals. An integer $i = [(m, n)]$ is positive if $m > n$. Then for integers i and j, we define $i < j$ if $j = i + p$ for some positive integer p. A rational $q = [(i, j)]$ is positive if either $i, j > 0$ or both $i, j < 0$. Again $p < q$ for rationals p, q if $q = p + r$ for some positive r.

We note that these orderings are not well-founded, since there is no least integer or rational number. The ordering on the rational numbers is studied more closely in the following section.

Exercises for Section 7.1

Exercise 7.1.1. Show that for every integers m and n, there is a natural number p such that (m, n) is equivalent to either $(0, p)$ or to $(p, 0)$.

Exercise 7.1.2. Show that the relation E_Z defined above is an equivalence relation, that is, reflexive, symmetric, and transitive.

Exercise 7.1.3. Show that the addition defined above on $\mathbb{Z}$ is well defined.

Exercise 7.1.4. Give a definition of multiplication for two integers, that is, define the operation $[(m_1, n_1)] \cdot [(m_2, n_2)]$.

Exercise 7.1.5. The set $\mathbb{N} = \omega$ of natural numbers has rank ω, that is, it lies in $V_{\omega+1}$. Find the rank of $\mathbb{Z}$ and of $\mathbb{Q}$ as defined above.

7.2 Dense Linear Orders

The basic step in the construction of the real line is the construction of the rational numbers. We show that the ordering of rational numbers is characterized up to isomorphism by two of its elementary properties: the density and the lack of endpoints.

Definition 7.2.1. A linear order $(K, \leq)$ is *dense* if for any elements $k_0 < k_1$ in K there is $k_2 \in K$ such that $k_0 < k_2 < k_1$.

Theorem 7.2.2. *Any two countable dense linear orders without endpoints are isomorphic.*

In other words, if $(K, \leq_K)$ and $(L, \leq_L)$ are countable dense linear orders without endpoints, then there is a bijective map $\phi \colon K \to L$ which is order-preserving: for any points $k_0, k_1 \in K$, $k_0 \leq_K k_1$ if and only if $\phi(k_0) \leq_L \phi(k_1)$ holds.

Proof. This is a so-called "back-and-forth argument" which appears in many other situations in mathematics. Use the countability assumption to fix enumerations $K = \{k_n \colon n \in \omega\}$ and $L = \{l_n \colon n \in \omega\}$. By recursion on n, build finite sets $a_n \subseteq K$ and $b_n \subseteq L$ and order-preserving bijective maps $\phi_n \colon a_n \to b_n$ such that $a_n \subseteq a_{n+1}$, $b_n \subseteq b_{n+1}$, $\phi_n \subseteq \phi_{n+1}$, $k_n \in a_{n+1}$, and $l_n \in b_{n+1}$. Once this is done, consider the map $\phi = \bigcup_n \phi_n \colon K \to L$. Since every element of K (and L, respectively) was placed in the domain (and range, respectively) of the map ϕ in its turn, it follows that $\text{Dmn}(\phi) = K$ and $Rng(\phi) = L$. Since each of the finite fragments ϕ_n of the map ϕ was order-preserving, so is ϕ. This will confirm the statement of the theorem.

To perform the construction, start with $a_0 = b_0 = \phi_0 = 0$. Suppose now that a_n, b_n, and ϕ_n have been constructed. We break the $n+1$ step into two similar smaller steps: First, we construct an order preserving bijection $\phi_{n+1}^0 \colon a_{n+1}^0 \to b_{n+1}^0$ extending ϕ_n such that $k_n \in \text{Dmn}(\phi_{n+1}^0)$. Then, an order-preserving bijection $\phi_{n+1} \colon a_{n+1} \to b_{n+1}$ extending ϕ_{n+1}^0 is constructed so that $l_n \in Rng(\phi_{n+1})$. This completes the construction at step $n+1$. The two smaller steps are symmetric, and we concentrate only on the first one.

The construction of ϕ_{n+1}^0 is performed depending on where the element $k_n \in K$ finds itself vis-a-vis a_n. There are several possibilities:

- if $k_n \in a_n$, then let $\phi_{n+1}^0 = \phi_n$;
- if the first item fails and k_n is $\leq_K$-smaller than all elements of a_n, then use the assumption that L has no smallest element to find

some $l \in L$ which is $\leq_L$-smaller than all elements of b_n, and let $a_{n+1} = a_n \cup \{k_n\}$, $b_{n+1} = b_n \cup \{l\}$, and $\phi_{n+1}^0 = \phi_n \cup \{(k_n, l)\}$;
- if the first item fails and k_n is $\leq_K$-larger than all elements of a_n, then act similarly as in the previous item;
- if all three previous items fail, then find a largest element $k \in a_n$ which is still $\leq_K$-smaller than k_n, and a smallest element $k' \in a_n$ which is still $\leq_K$-larger than k_n, use the density assumption to find an element $l \in L$ such that $\phi_n(k) <_L l < \phi_n(k')$, and let $a_{n+1} = a_n \cup \{k_n\}$, $b_{n+1} = b_n \cup \{l\}$, and $\phi_{n+1}^0 = \phi_n \cup \{(k_n, l)\}$.

Since there are no other options, this successfully completes the induction step and the proof. $\qquad\square$

Theorem 7.2.2 makes it possible to characterize the rationals $(\mathbb{Q}, \leq)$ with their linear ordering as the unique (up to isomorphism) countable dense linear order without endpoints.

Exercises for Section 7.2

Exercise 7.2.1. Show that the density requirement in Theorem 7.2.2 is necessary. That is, find two countable infinite linear orders without endpoints which are not isomorphic.

Exercise 7.2.2. Call a graph (V, E) a *random graph* if for every pair a, b of disjoint finite subsets of V there is a vertex $v \in V$ which is connected to all elements of a and to no elements of b. Use a back-and-forth argument to show that any two countably infinite random graphs are isomorphic.

7.3 Complete Orders

The real line is constructed as a *completion* of the linear ordering of rational numbers. This allows us to speak about concrete objects such as $\sqrt{2}$ which cannot be rational numbers. It also enables the abstract definitions of differentiation and integration. We show that there is a general operation of completion of dense linear orders which is defined up to isomorphism. To formulate the concept precisely, we need a couple of definitions. The reader is referred to Chapter 2 for basic concepts about linear orders.

Definition 7.3.1. Let $(L, \leq)$ be a linear ordering.

1. An *open interval* of L is a set of the form $\{k \in L : l_0 < k < l_1\}$ for some elements $l_0 < l_1$ of L. The sets of the form $\{k \in L : k < l\}$ and $\{k \in L : l < k\}$ are also considered to be open intervals of L and so is L itself.
2. A subset $K \subseteq L$ is *dense in* L if K has nonempty intersection with every nonempty open interval of L.
3. The ordering L is *complete* if every non-empty bounded set $K \subseteq L$ has a supremum.

Definition 7.3.2. Let $(K, \leq_K)$ be a dense linear order without endpoints. A *completion* of K is a complete dense linear order $(L, \leq)$ without endpoints together with an order preserving map $\phi \colon K \to L$ such that $Rng(\phi) \subseteq L$ is dense in L.

The following definition is also needed.

Definition 7.3.3. Let $(L, \leq)$ be a linear order and let $A \subseteq L$.

1. A is downwards closed if $k_0 \leq_K k_1$ and $k_1 \in A$ implies $k_0 \in A$;
2. A is upwards closed if $k_0 \geq_K k_1$ and $k_1 \in A$ implies $k_0 \in A$.

Theorem 7.3.4. *Let $(K, \leq_K)$ be a dense linear order without endpoints. Then $(K, \leq_K)$ has a completion which is in addition unique up to isomorphism.*

The last sentence needs clarification; the precise statement is the following. Suppose that K_0, K_1 are dense linear orders without endpoints and $\psi \colon K_0 \to K_1$ is an order isomorphism. Suppose that $\phi_0 \colon K_0 \to L_0$ and $\phi_1 \colon K_1 \to L_1$ are completions of K_0, K_1 respectively. Then there is an order isomorphism $\chi \colon L_0 \to L_1$ such that $\phi_1 \circ \psi = \chi \circ \phi_0$.

Proof. We present a classical construction due to Dedekind, a German mathematician who was active in the second half of 19th century. A pair (A, B) is called a *Dedekind cut* if A, B are non-empty disjoint subsets of K such that $A \cup B = K$, A is downwards closed ($k_0 \leq_K k_1$ and $k_1 \in A$ implies $k_0 \in A$), and A has no maximal element (for every $k_0 \in A$ there is $k_1 \in A$ such that $k_0 <_K k_1$). If K is the set of rational numbers, then, for example, we might have $A = \{x : x < 0 \ \lor \ x^2 < 2\}$ and $B = \{x : x > 0 \ \land \ x^2 > 2\}$.

This Dedekind cut would represent the real number $\sqrt{2}$. Let L be the set of all Dedekind cuts, ordered by inclusion in the first coordinate: $(A_0, B_0) \leq_L (A_1, B_1)$ if $A_0 \subseteq A_1$. We show that this is the completion of the linear order K in a series of claims.

Claim 7.3.5. *The ordering on L is linear.*

Proof. Let (A_0, B_0) and (A_1, B_1) be Dedekind cuts. Suppose towards contradiction that neither $A_0 \subseteq A_1$ nor $A_1 \subseteq A_0$ holds. Then, there must be elements $k_0 \in A_0 \setminus A_1$ and $k_1 \in A_1 \setminus A_0$. By the linearity of the ordering $\leq_K$, either $k_0 \leq_K k_1$ or $k_1 \leq_K k_0$ must occur, and both cases lead to contradiction: If $k_0 \leq_K k_1$, then $k_0 \in A_1$ by the downward closure of A_1, and this contradicts the choice of k_0. If, on the other hand, $k_1 \leq_K k_0$, then $k_1 \in A_0$ by the downward closure of A_0, and this contradicts the choice of k_1. $\square$

Claim 7.3.6. *The ordering on L is complete.*

Proof. Let $J \subseteq L$ be a nonempty bounded set; we must produce its supremum. Let $C = \bigcup \{ A \subseteq K : \text{ for some } B \subseteq K, (A, B) \in J \}$ and let $D = K \setminus C$. We show that the pair (C, D) is a Dedekind cut and a supremum of the set J.

First, of all, the set $C \subseteq K$ is downward closed and has no maximal element because it is a union of downward closed sets without a maximal element. Also, the set D is non-empty since the set $J \subseteq L$ is bounded: If (C', D') is an upper bound of J in L, then $C \subseteq C'$ must hold, and so $D' \subseteq D$ must hold. Thus, the pair (C, D) is a Dedekind cut. To show that (C, D) is the supremum of J, suppose that (C', D') is an upper bound of J and use the definition of the ordering on L to observe once again that $C \subseteq C'$. This means exactly that $(C, D) \leq (C', D')$ as required. $\square$

For each point $k_0 \in K$, let $\phi(k_0) = (A, B)$, where $A = \{ k_1 \in K : k_1 <_K k_0 \}$ and $B = \{ k_1 \in K : k_0 \leq_K k_1 \}$.

Claim 7.3.7. *The map $\phi \colon K \to L$ is an order-preserving injection. Moreover, $Rng(\phi) \subseteq L$ is dense.*

Proof. The first sentence is immediate. For the density of $Rng(\phi)$ in L, assume that (A_0, B_0) and (A_1, B_1) are two Dedekind cuts, $A_0 \subseteq A_1$, and $A_0 \neq A_1$; we must produce $k \in K$ such that $\phi(k)$ is strictly between the two cuts. Let $k_1 \in A_1 \setminus A_0$ be an arbitrary element. Since

A_1 has no maximum, there is an element $k_0 \in A_1$ strictly larger than k_1; we claim that k_0 works. To this end, we must prove that the set $C = \{k \in K \colon k <_K k_0\}$ is strictly between A_0 and A_1. Certainly, $C \subseteq A_1$ since $k_0 \in A_1$ and A_1 is closed downwards; also, $C \neq A_1$ since $k_0 \in A_1 \setminus C$. To see that $A_0 \subseteq C$, note that all elements of A_0 must be below k_0 since $k_0 \notin A_0$ and A_0 is downward closed. Finally, $A_0 \neq C$ since $k_1 \in C \setminus A_0$. □

The final point is the uniqueness of the completion. Suppose that K_0, K_1 are densely linearly ordered sets without endpoints and $\psi \colon K_0 \to K_1$ is an order isomorphism. Let L_1 be the completion of K_1 via Dedekind cuts, and let L_0 be an arbitrary completion of K_0. For every element $l \in L_0$, let $A_l = \{\psi(k) \colon k \in K_0 \text{ and } k < l\} \subseteq K_1$, and let $B_l = K_1 \setminus A_l$. It is not difficult to verify that (A_l, B_l) is a Dedekind cut in K_1, and the map $\chi \colon l \mapsto (A_l, B_l)$ is the required order isomorphism of L_0 and L_1. □

The work done so far allows the characterization of the real line as the unique (up to isomorphism) complete dense linear order without endpoints and containing a countable dense subset. Note that the countable dense subset has to be densely ordered and does not have any endpoints and therefore must be order-isomorphic to the rationals by Theorem 7.2.2. The ordering $(\mathbb{R}, \leq)$ is then order-isomorphic to the completion of the rationals in the sense of Theorem 7.3.4.

The description of the real line in these terms leads to a deep question in pure set theory. The question (known as *Souslin's problem*) [10] suggests an alternative characterization of the real line; it was first published in 1920. Mikhail Souslin was a graduate student of mathematics in Moscow who died young in the chaos of Russian Civil War.

Question 7.3.8 (Souslin). Let $(L, \leq)$ be a complete dense linear order without endpoints. Are the following equivalent?

1. $(L, \leq)$ is isomorphic to the real line.
2. Every collection of pairwise disjoint non-empty intervals of L is countable.

Note that if L contains a countable dense set $K \subseteq L$ (as is the case with the real line), then indeed every collection A of pairwise disjoint non-empty intervals of L is countable. To see this, consider a function which assigns each interval in A one of its elements in the dense set K. Such a function must be an injection, since the intervals in A are assumed to be pairwise disjoint. It follows that A can be injected into the countable set K and therefore is itself countable. This proves the implication (1)→(2) in the above question. But why should the opposite implication hold? The question was shown to be irresolvable within the framework of ZFC set theory in 1971.

We conclude this section with a basic cardinality computation.

Theorem 7.3.9. *The real line has the same cardinality as $\{0,1\}^\omega$.*

Proof. By the Schröder–Bernstein theorem, we need to produce an injection from $\mathbb{R}$ to the powerset of a countable set, and an injection from $\mathcal{P}(\omega)$ to $\mathbb{R}$. For the former task, let $f \colon \mathbb{R} \to \mathcal{P}(\mathbb{Q})$ be the map assigning each real number r the set $\{q \in \mathbb{Q} \colon q < r\}$. Since the rational numbers are dense in the real line, this is indeed an injection.

For the latter task, we define an injection from $\{0,1\}^\omega$ into the real interval $[0,1]$ by $g(x) = \sum_{n=0}^\infty 2x(n)3^{-n-1}$. For example, $g(1,0,1,0,\dots) = 2/3 + 2/27 + \cdots = 2\sum_{n=0}^\infty 3^{-2n} = 3/4$. To see that this is an injection, suppose that $x \neq y \in \{0,1\}^\omega$ and let n be the least such that $x(n) \neq y(n)$; without loss of generality say that $x(n) = 0$ and $y(n) = 1$. Then $\sum_{i=0}^n 2y(i)3^{-i-1} = 2 \cdot 3^{-n-1} + \sum_{i=1}^n 2x(i)3^{-i-1}$. Now, $\sum_{i=n+1}^\infty 2 \cdot 3^{-i-1} = 3^{-n-1}$ so that

$$g(x) \leq 3^{-n-1} + \sum_{i=0}^n 2x(i)3^{-i-1} < \sum_{i=0}^n 2y(i)3^{-i-1} \leq g(y).$$

$\square$

The range of the function f is commonly known as the *Cantor middle third set*.

Exercises for Section 7.3

Exercise 7.3.1. Show that if (A,B) is a Dedekind cut in a linear order $(L, \leq)$, then B is upwards closed.

Exercise 7.3.2. Show that if C is the union of a family of downwards closed sets, then C is downwards closed.

Exercise 7.3.3. Show that if C is the union of a family of sets, each of which has no maximal element, then C has no maximal element.

Exercise 7.3.4. Let K be a dense linear order without endpoints. Show that the completion of K has cardinality at most that of $\mathcal{P}(K)$. Is it possible that the completion of K has the same cardinality as K?

Exercise 7.3.5. Let $(K, \leq)$ be a complete linear order, and let $(I_n : n \in \omega)$ be a nested sequence of bounded closed intervals in K; that is, $I_{n+1} \subseteq I_n$ holds for each $n \in \omega$. Show that $\bigcap_n I_n \neq 0$.

Exercise 7.3.6. Let $(K, \leq)$ be a complete linear order. Show that every set $A \subseteq L$ bounded from below has an infinimum; that is, there is an element k such that for all $l \in A$, $k \leq l$, and k is the largest such element.

Exercise 7.3.7. Show that there is no order-preserving injection from ω_1 to $\mathbb{R}$.

Exercise 7.3.8. Let $(K, \leq_K)$ be a dense linear order without endpoints and $\phi \colon (K, \leq_K) \to (L, \leq_L)$ be its completion. Show that for any order-preserving map $\psi \colon (K, \leq_K) \to (M, \leq_M)$ such that $Rng(\psi) \subseteq M$ is dense in M there is an order-preserving map $\chi \colon (M, \leq_M) \to (L, \leq_L)$ such that $\phi = \chi \circ \psi$. Thus, the completion of K is in a sense the largest linear order in which K is dense.

7.4 Countable and Uncountable Sets of Reals

In this section, we examine the structure of infinite sets of real numbers. The question of the cardinality of infinite sets of reals led to the so-called *Continuum Hypothesis*, which was one of the major themes of 20th century mathematics and, along with the *Generalized Continuum Hypothesis*, continues to be of great interest. The Continuum Hypothesis states that every set of reals is either countable or has the cardinality of the real numbers (the *continuum*). We see that the hypothesis holds for *closed* sets.

Our analysis depends on the Cantor–Bendixson theorem, given in the following, which is a typical application of transfinite recursion to mathematical analysis. We need some background first. Recall that a *basic open* set of reals is an interval (p, r) with rational endpoints, not including the endpoints. It is important to note that there are countably many such intervals, so they can be enumerated as $\{I_n : n < \omega\}$. An open set of reals is one which is obtained as a union of some collection of basic open sets, and a closed set is one whose complement is open. A set U is said to be a *neighborhood* of a point $x \in U$ if U is open. A point $x \in A \subseteq \mathbb{R}$ is *isolated* in the set A if there is an open interval which contains x and no other points of the set A. A subset P of $\mathbb{R}$ is *bounded* if there is some b such that $|x| < b$ for all $x \in P$; P is said to be *compact* if it is closed and bounded.

A point p is said to be a *point of condensation* of a set P if every neighborhood of p contains uncountably many points of P; p is said to be a *point of accumulation* of P if every neighborhood of p contains at least one point of P different from p; it is an exercise to show that this is equivalent to each neighborhood of p having infinitely many points of P. Note that an element of P which is not an accumulation point is isolated in P. A closed set without isolated points is said to be *perfect*.

Proposition 7.4.1. *Any infinite subset of $[0,1]$ (or any compact interval) must have a point of accumulation.*

Proof. Here is a sketch: Let P be an infinite subset of $[0,1]$. Observe that either $P \cap [0, \frac{1}{2}]$ or $P \cap [\frac{1}{2}, 1]$ must be infinite and let $I_1 = [a_1, b_1]$ be one of the intervals such that $P \cap I_0$ is infinite. Now, divide I_1 in half and let $I_2 \subseteq I_1$ be an interval $[a_2, b_2]$ of diameter $\frac{1}{4}$ such that $P \cap I_2$ is infinite. Note that $a_1 \leq a_2 < b_2 \leq b_1$. Continuing in this way, we obtain a decreasing chain $[0,1] \supseteq I_1 \supseteq I_2 \supseteq \cdots$ of intervals $I_n = [a_n, b_n]$ with $a_n \leq a_{n+1} < b_{n+1} \leq b_n$ and $b_n - a_n = 2^{-n}$ such that $P \cap I_n$ is infinite for each n. Then there is a real p such that $\bigcap_n I_n = \{p\}$ and this p is an accumulation point of P. To see that p exists, consider the non-decreasing sequence $a_1 \leq a_2 \leq \cdots$ which is bounded above by 1. Then $\lim_n a_n = p$ must exists by the completeness of $\mathbb{R}$. Since $b_n - a_n = 2^{-n}$, it follows that $p = \lim_n b_n$ as well. To confirm that p is an accumulation point, let $I = (p - \epsilon, p + \epsilon)$ be any neighborhood of p and let n be large enough so $2^{-n} < \epsilon$. Since $p \in I_n$ and I_n has diameter 2^{-n}, it follows that $I_n \subseteq I$ and therefore $P \cap I$ is infinite. $\square$

We leave as an exercise that any uncountable set of reals must have a point of condensation.

The Cantor–Bendixson derivative $D(X)$ of a closed set X is defined to be $X \setminus \{\text{isolated points of } X\}$. For example, if $X =$

$\{0\} \cup \{2^{-n} : n \in \omega\}$, then $D(X) = \{0\}$. The derivative can be iterated through the ordinals as follows:

Definition 7.4.2. For any closed set X, any ordinal α and any limit ordinal λ, let

1. $D^{\alpha+1}(X) = D(D^\alpha(X))$;
2. $D^\lambda(X) = \bigcap_{\beta < \lambda} D^\beta(X)$;
3. the Cantor–Bendixson rank of X is the least α such that $D^{\alpha+1}(X) = D^\alpha(X)$.

Example 7.4.3. Let $C = \{0\} \cup \{\frac{1}{n} : n \in \omega\}$. Then $D(C) = \{0\}$ and $D^2(C) = \emptyset = D^3(C)$. Let $K = C \cup \{\frac{1}{n} + \frac{1}{n \cdot 2^m} : m, n \in \omega\}$. Then $D(K) = C$ so that $D^2(K) = \{0\}$ and $D^3(K) = \emptyset = D^4(K)$. Thus, C has Cantor–Bendixson rank 2 and K has Cantor–Bendixson rank 3.

Recall from Example 5.4.6 the closed, well-ordered set $P_0 = \{n + 1 - 2^{-k} : n, k \in \omega\} = \{0, \frac{1}{2}, \frac{3}{4}, \ldots, 1, \frac{3}{2}, \frac{7}{4}, \ldots, 2, \ldots\}$. Then $D(P_0) = \{1, 2, \ldots\}$ and $D^2(P_0) = D(P_0)$. Let $\Phi : P \to \omega^2 + 1$ map each element of P to its position in the order type. We observe that $\Phi[D(P)] = \{\omega, \omega \cdot 2, \ldots\}$, that is, the set of limit ordinals in $\Phi[P]$. Then $\Phi(D^2(P)] = \emptyset$, that is, the (empty) set of ordinals in ω^2 which are limits of limit ordinals.

It is easy to see that for any closed set X, $D(X)$ is also closed. That is, if $x \notin D(X)$, then x is in some interval I such that $X \cap I \subseteq \{x\}$ so that $D(X) \cap I = \emptyset$.

Lemma 7.4.4. *Show that for any closed set C, $C \setminus D(C)$ is countable.*

The proof is left as an exercise.

Lemma 7.4.5. *For any closed set C of reals and any ordinals $\beta < \alpha$, $D^\alpha(C)$ is closed and $D^\alpha(C) \subseteq D^\beta(C)$.*

Proof. We proceed by transfinite induction on α. At limit stage α, the construction takes an intersection of a collection of closed sets, which then must be closed and smaller than all sets in the intersection. At the successor stage, $D^{\alpha+1}(C) \subseteq D^\alpha(C)$ certainly holds. To prove that $D^{\alpha+1}(C)$ is closed, for every point $x \in D^\alpha(C) \setminus D^{\alpha+1}(C)$ pick an open neighborhood O_x containing only x and no other elements of $D^\alpha(C)$. Then $D^{\alpha+1}(C) = D^\alpha(C) \setminus \bigcup_x O_x$, and as a difference of a closed set and an open set, the set $D^\alpha(C)$ is closed. □

The following result will be useful.

Proposition 7.4.6. *Any properly decreasing chain of closed sets must be countable.*

Proof. Let $\{C_\alpha : \alpha < \beta\}$ be a sequence of closed sets such that $C_\alpha \subseteq C_\gamma$ whenever $\gamma < \alpha$. Let α be given so that $\alpha + 1 < \beta$. Since the chain is properly decreasing, we must have some $x \in C_\alpha \setminus C_{\alpha+1}$. Since $x \notin C_{\alpha+1}$, which is a closed set, there is some rational interval I_n such that $I_n \cap C_{\alpha+1} = \emptyset$. On the other hand, $x \in C_\alpha$ so that $I_n \cap C_\alpha \neq \emptyset$. Now, define a function f from $\beta \to \omega$ by letting $f(\alpha)$ be the least n such that $I_n \cap C_{\alpha+1} = \emptyset$ but $I_n \cap C_\alpha \neq \emptyset$. If $\gamma < \alpha < \beta$, then $C_\alpha \subseteq C_{\gamma+1}$ so that $I_{f(\gamma)} \cap C_\alpha = \emptyset$, but $I_{f(\alpha)} \cap C_\alpha \neq \emptyset$ so that $f(\gamma) \neq f(\alpha)$. Thus, f is an injection from β into ω and it follows that β is a countable ordinal. $\square$

Theorem 7.4.7 (Cantor–Bendixson [2]). *Every closed set of reals can be written as a disjoint union of a countable set and a perfect closed set.*

In fact, the decomposition is unique, as we show later.

Proof. Let $C \subseteq \mathbb{R}$ be a closed set of reals and define as above the sequence $D^\alpha(C)$ for all ordinals α. It follows from Proposition 7.4.6 that there exists a countable ordinal β such that the iteration of the Cantor–Bendixson derivative on C stabilizes in the sense that $D^{\beta+1}(C) = D^\beta(C)$. Let β be the smallest such ordinal, and let $K = C \setminus D^\beta(C)$. $D^\beta(C)$ is sometimes called the *perfect kernel* of C. We show that $C = D^\beta(C) \cup K$ is the desired decomposition of C into a perfect closed set and a countable set.

First of all, it is clear that the set $D^\beta(C)$ has no isolated points as $D^{\beta+1}(C) = D^\beta(C) \setminus \{$isolated points of $D^\beta(C)\}$ by the recursive definition, and $D^{\beta+1}(C) = D^\beta(C)$.

For each ordinal α, $D^\alpha(C) \setminus D^{\alpha+1}(C)$ is countable by Lemma 7.4.4. Then by Theorem 4.5.2 (which uses the Axiom of Choice), we can see that $C \setminus D^\beta(C) = \bigcup_{\alpha<\beta} D^\alpha(C) \setminus D^{\alpha+1}(C)$ is countable. $\square$

To obtain the uniqueness of the decomposition given by the Cantor–Bendixson theorem, we need to consider some further results. We want to show that any countable set of reals must have an isolated point so that a non-empty perfect set is uncountable. In fact, we next see that a non-empty perfect set must have cardinality of the continuum.

Theorem 7.4.8. *If $P \subseteq \mathbb{R}$ is a non-empty perfect closed set, then $|P| = |\mathbb{R}|$.*

Proof. It suffices to find an injection from $\mathbb{R}$ into P. We do this as follows. Since P is non-empty, it contains a point p_0; since P is perfect, it must contain infinitely many other points. So, it contains at least a second point p_1. Now, choose open intervals J_0 and J_1, having disjoint closures, with $p_0 \in J_0$ and $p_1 \in J_1$. It is easy to see that $P \cap J_0$ and $P \cap J_1$ have no isolated points. (See Exercise 7.4.11.) Repeat this process to find elements $p_{00}, p_{01}, p_{10}, p_{11}$ and intervals $J_{ij} \subseteq J_i$ with $p_{ij} \in P \cap J_{ij}$ so that $P \cap J_{ij}$ has no isolated points. Continuing recursively, we obtain, for every $n \in \omega$, a set of points $\{p_\sigma : \sigma \in \{0,1\}^n\}$ and a family $\{J_\sigma : \sigma \in \{0,1\}^n\}$ of open intervals intervals such that, for all $\sigma \in \{0,1\}^n$,

1. the closures of the intervals are disjoint,
2. $p_\sigma \in P \cap J_\sigma$,
3. for $k = 0, 1$, $J_{\sigma k} \subseteq J_\sigma$,
4. J_σ has diameter $\leq 2^{-n}$.

Now, define an injection from $\{0,1\}^\omega$ into P by letting $F(X) = \lim_{n \to \infty} p_{X \restriction n}$. This limit exists since the intervals are shrinking to zero in size and $F(X)$ belongs to P since P is closed. It follows that $\bigcap_n J_{X \restriction n} = \{F(X)\}$. If $X \neq Y$, then for some n, $X \restriction n \neq Y \restriction n$ so that $F(X) \neq F(Y)$. $\qquad \square$

Corollary 7.4.9. *If P is a countably infinite closed set, then P has an isolated point.*

Proof. We prove the contrapositive. Suppose that P has no isolated point. Then by definition P is perfect. Thus, P is uncountable by Theorem 7.4.8. $\qquad \square$

Corollary 7.4.10. *If P is a perfect closed set, then every element of P is a point of condensation of P.*

Proof. Let $p \in P$ and let J be any interval containing P. Then $P \cap J$ is also perfect, by Exercise 7.4.11. Thus, $P \cap J$ is uncountable by Theorem 7.4.8. $\qquad \square$

Corollary 7.4.11. *For any closed set C, the Cantor–Bendixson decomposition of C into a union $K \cup P$ of a countable set K and a perfect set P is unique.*

Proof. Suppose that $C = K_1 \cup P_1 = K_2 \cup P_2$, where K_1 and K_2 are countable and P_1 and P_2 are perfect closed sets, and $K_i \cap P_i = \emptyset$ for $i = 1, 2$. Suppose by way of contradiction that $K_1 \cap P_2 \neq \emptyset$ and let $p \in K_1 \cap P_2$. Let J be any interval containing p. Since $p \in P_2$, p is a point of condensation of P_2, so there are uncountably many elements of P_2 in J. Since K_1 is countable, there must be uncountably many elements of P_1 in J. Thus, every interval J containing p has at least *one* element of P_1. Since P_1 is closed, it follows that $p \in P_1$, violating the assumption that $K_1 \cap P_1 = \emptyset$. A similar argument shows that $K_2 \cap P_1 = \emptyset$, and it follows that $K_1 = K_2$ and $P_1 = P_2$. $\square$

Theorem 7.4.12. *For every closed set $C \subseteq \mathbb{R}$, either C is countable or it has the cardinality of continuum.*

Proof. Express C as a union $C = D \cup E$ of a countable set D and a perfect set E. If the set E is empty, then the set C is equal to D and therefore countable. If, on the other hand, the set E is non-empty, then it has cardinality continuum by Theorem 7.4.8. Thus, C is sandwiched between two sets E and $\mathbb{R}$ of cardinality continuum and it must have that same cardinality itself. $\square$

Exercises for Section 7.4

Exercise 7.4.1. Provide an example of a closed set $C \subseteq \mathbb{R}$ which has Cantor–Bendixson rank ω.

Exercise 7.4.2. For any set $X \subseteq \mathbb{R}$, define $\Gamma(X) = \mathbb{R} \setminus D(\mathbb{R} \setminus X)$. Show that Γ is a monotone operator.

Exercise 7.4.3. Show that if P and Q are closed, then $D(P \cap Q) \subseteq D(P) \cap D(Q)$. Give an example to show that equality does not always hold.

Exercise 7.4.4. Show that if P and Q are closed, then $D(P \cup Q) = D(P) \cup D(Q)$.

Exercise 7.4.5. Show that x is an accumulation point of P if and only if there is an infinite sequence $\{y_n : n \in \omega\}$ of points y_n such that $\lim_n y_n = x$ with $y_n \neq x$ and $y_n \in P$ for each n.

Exercise 7.4.6. Show that p is a point of accumulation of P if and only if every neighborhood of p contains infinitely many points of P.

Exercise 7.4.7. Show that if P is a compact, well-ordered set of real numbers, then P has a maximal element.

Exercise 7.4.8. Let $P = \{1\} \cup \{1 - 2^{-n} + 2^{-n-1}(1 - 2^{-k}) : n, m \in \omega\}$ and let Φ be the order isomorphism taking P to its order type $\omega^2 + 1$. Find $D(P)$ and $D^2(P)$. Show that $\Phi[D(P)]$ is the set of limit ordinals in $\omega^2 + 1$ and that $\Phi(D^2(P)]$ is the set of ordinals which are limits of limits.

Exercise 7.4.9. Show that for any closed, well-ordered set P of reals, $x \in D(P)$ if and only if $\Phi(x)$ is a limit ordinal, where $\Phi : P \to \alpha$ is the canonical isomorphism from P to its order type α.

Exercise 7.4.10. Show that if P is a compact subset of $\mathbb{R}$ and $D(P) = \emptyset$, then P is finite.

Exercise 7.4.11. Show that if P is a set with no isolated points (not necessarily closed) and I is an open interval, then $P \cap I$ has no isolated points.

Exercise 7.4.12. Show that for any closed set C, $C \setminus D(C)$ is countable. *Hint*: Define a mapping from $C \setminus D(C)$ into ω using the countable basis $I_0, I_1, \ldots$.

Exercise 7.4.13. Show that if $P \subseteq \mathbb{R}$ is uncountable, then it has a point of condensation. *Hint*: Prove by contrapositive and use the countable basis of $\mathbb{R}$.

Exercise 7.4.14. Modify the proof of Lemma 7.4.5 to show that the set $K = C \setminus D^\beta(C)$ is countable in the proof of the Cantor–Bendixson theorem. That is, construct an injection F from K to ω so that if $x \in D^\alpha(C) \setminus D^{\alpha+1}(C)$, then $I_{F(x)}$ contains only x and no other elements of $D^\alpha(C)$. This argument avoids the Axiom of Choice.

Exercise 7.4.15. Let $(K, \leq)$ be a linear ordering. Show that $K = L_0 \cup L_1$, where L_0 is well-ordering and L_1 does not have a smallest element.

7.5 Topological Spaces

Many objects in mathematics are equipped with a structure that makes it possible to speak about continuous functions from one object to another — a topology.

Definition 7.5.1. A *topological space* is a pair (X, T), where X is a non-empty set and $T \subseteq \mathcal{P}(X)$ is a collection of subsets of X containing $\emptyset$ and X and closed under finite intersections and arbitrary unions. The collection T is a *topology* and its elements are referred to as *open sets*.

Definition 7.5.2. Suppose that (X, T) and (Y, U) are two topological spaces. A map $f : X \to Y$ is *continuous* if the f-preimages of open subsets of Y are open in X; that is, $f^{-1}[O]$ is open in X for every open $O \subseteq Y$. The map f is a *homeomorphism* if it is a bijection and both f and f^{-1} are continuous maps. A family $\mathcal{B}$ of subsets of X is said to be a *basis* for T if, for every point $x \in X$ and every open set U such that $x \in U$, there is a set $B \in \mathcal{B}$ such that $x \in B$ and $B \subseteq U$. A family $\mathcal{S}$ of subsets of X is said to be a *subbasis* for T if the family of all finite intersections of sets from $\mathcal{S}$ is a basis.

Before we turn to examples, it is useful to note that most topologies are generated from subbases in the following way:

Definition 7.5.3. Let X be a set and $\mathcal{S} \subseteq \mathcal{P}(X)$ be a subbasis. The *topology generated by* $\mathcal{S}$ is the set $T = \{O \subseteq X : O = \bigcup B$ for some set B consisting of finite intersections of elements of $\mathcal{S}\} \cup \{\emptyset, X\}$.

Proposition 7.5.4. *Whenever X is a set and $\mathcal{S} \subseteq \mathcal{P}(X)$, the collection T above is in fact a topology on X.*

Proof. Clearly, $\emptyset, X \in T$ by the definition of T. We have to prove that T is closed under arbitrary unions and finite intersections.

The closure under arbitrary unions is immediate. If $U \subseteq T$ is any set, we must show that $\bigcup U \in T$. Let

$$B = \left\{ P \subseteq X : (\exists O_1, \ldots O_n \in \mathcal{S}) \, P = \bigcap_{i=1}^{n} O_i \wedge (\exists O \in U) \, P \subseteq O \right\}.$$

It is not difficult to check that $\bigcup B = \bigcup U$ and so $\bigcup U \in T$ as required.

Now, we must show that T is closed under finite intersections. If $U \subseteq T$ is a finite set, we must show that $\bigcap U \in T$. Let

$$B = \left\{ P \subseteq X : (\exists O_1, \ldots O_n \in \mathcal{S}) \, P = \bigcap_{i=1}^{n} O_i \wedge P \subseteq \bigcap U \right\}.$$

We show that $\bigcup B = \bigcap U$; this proves that $\bigcap U \in T$ as required. For the $\bigcup B \subseteq \bigcap U$ inclusion, note that B by definition consists of sets which are subsets of $\bigcap U$. For the $\bigcap U \subseteq \bigcup B$ inclusion, let $x \in \bigcap U$ be an arbitrary point. Since $U \subseteq T$, for every set $O \in U$, there is a set $P_O \subseteq O$ which is an intersection of finitely many elements of $\mathcal{S}$ and contains the point x. Since U is finite, the set $\bigcap_{O \in U} P_O$ is an intersection of finitely many elements of $\mathcal{S}$, it is in B, and it contains the point x. Therefore, $x \in \bigcup B$. $\qquad\square$

Example 7.5.5. The *discrete topology* on a set X is $T = \mathcal{P}(X)$. In other words, every set is open in the discrete topology.

Example 7.5.6. If $(L, \leq)$ is a linear ordering, the *order topology* is generated by the basis consisting of all sets of the form (p, q), where $p < q$ are elements of L and (p, q) is the *open interval* $\{r : p < r < q\}$.

Example 7.5.7. The *Cantor space* is the set $2^{\mathbb{N}} = \{0, 1\}^{\mathbb{N}}$, and the set of infinite binary sequences is equipped with the topology generated by the subbasis consisting of all sets of the form $\{f \in 2^{\mathbb{N}} : f(n) = b\}$, where $n \in \omega$ and $b \in \{0, 1\}$. For any finite sequence $\sigma = (i_0, \ldots, i_{n-1})$, let

$$[\![\sigma]\!] = \{x \in \{0, 1\}^{\mathbb{N}} : (\forall j < n) \, x(j) = i_j\};$$

these are a basis for the topology and are called *intervals*.

Example 7.5.8. The *Baire space* is the set $\mathbb{N}^{\mathbb{N}}$, and the set of infinite sequences of natural numbers is equipped with the topology generated by the subbasis consisting of all sets of the form $\{f \in \mathbb{N}^{\mathbb{N}} : f(n) = m\}$, where $n, m \in \omega$. Again, the intervals $[\![\sigma]\!]$ are a basis for the topology, where the definition of $[\![\sigma]\!]$ is suitably modified for $\mathbb{N}^{\mathbb{N}}$.

The problem of finding a member of a closed set in the Baire space or Cantor space is very important in effective mathematics and in proof theory. Recall that a tree T in ω^* is a set which is closed under initial segments. Also, $x \in \omega^\omega$ is an infinite path through T if and only if $x \upharpoonright n \in T$ for all n and that $[T]$ is the set of infinite paths through T.

In the Cantor space, every infinite tree has an infinite path. This follows from the following König's Lemma. But this is not true in the Baire space.

Example 7.5.9. Let $T = \{n0^i : i < n\}$ as a subset of ω^* (see Section 2.6). Then T is an infinite tree and also has arbitrarily long finite paths. But it has no infinite path.

Lemma 7.5.10 (König's Lemma). *If $T \subseteq \omega^*$ is infinite and finite-branching, then T has an infinite path.*

Proof. Let $T \subseteq \omega^*$ be an infinite, finite-branching tree. Now, let $S \subseteq T$ be the set of nodes $\sigma \in T$ such that T contains infinitely many extensions of σ. The empty node $\epsilon \in S$ since T is infinite. It follows from the Pigeonhole Principle (Corollary 4.2.7) that if $\sigma \in S$, then at least one successor of σ belongs to S. That is, since T is finite-branching, the infinite set of extensions of σ is partitioned into finitely many sets which are the extensions of the successors of σ. We now define an element x of $[T]$ as follows. Let $x(0)$ be the least i such that $[\![i]\!] \in S$. For any n, let $x(n)$ be the least i such that $(x(0), x(1), \ldots, x(n-1), i) \in T$. $\qquad \square$

In set theory and logic, we often refer to elements of the Cantor space, or the Baire space, as reals. What justifies this convention? For the Cantor space, consider the function mapping $x \in \{0,1\}^\omega$ into the real interval $[0,1]$, defined by $F(x) = \sum_n x(n)/2^{n+1}$; x is said to be the dyadic expansion of the real $F(x)$. This is a continuous map from $\{0,1\}^\omega$ onto $[0,1]$, but it is not quite one-to-one. That is, each dyadic rational has two representations. For example, $3/4 = 1/2 + 1/4$ has representation $(1,1,0,0,\ldots)$ but also we have $3/4 = 1/2 + 1/8/ + 1/16 + \ldots$ and thus has representation $(1,0,1,1,\ldots)$. However, the set of dyadic rationals is countable, so for many situations it will be all right to imagine that elements of $\{0,1\}^\omega$ are real numbers.

For the Baire space, we have the very interesting continued fraction representation, that is,

$$G(x) = \cfrac{1}{x(0) + \cfrac{1}{x(1) + 1/\ldots}}.$$

It can be seen that this is a continuous isomorphism of ω^ω onto the space of irrational numbers in the interval $[0, 1]$. Since there are only countably many rational numbers, this makes ω^ω a reasonable version of the real numbers.

On the other hand, there is a big difference between the topologies on 2^ω and ω^ω, on the one hand, and $[0, 1]$ and $\mathbb{R}$, on the other. The real line is connected, meaning that there are no sets, other than $\mathbb{R}$ and $\emptyset$, which are both open and closed (clopen). But in the Cantor space, every basic open set is also a closed set. Thus, the Cantor space is said to be totally disconnected, that is, for any point $x \neq y$, there is a clopen set which contains x but does not contain y. This is also true in the Baire space. In the space of irrationals, notice that the open interval $(1/2, 3/4)$ equals the closed interval $[1/2, 3/4]$, since $1/2$ and $3/4$ are not in the space.

Closed sets have an interesting and useful characterization in the Cantor space (as well as the Baire space) in terms of trees.

Proposition 7.5.11. *A subset K of $\mathbb{N}^\mathbb{N}$ is closed if and only if there is a tree T such that $[T] = K$.*

Proof. Suppose first that K is closed and let $T = \{\sigma \in \omega^* : K \cap [\![\sigma]\!] \neq \emptyset\}$. It is left to the reader to check that T is a tree. We claim that $K = [T]$. If $x \in K$, then for any n, $x \in K \cap [\![x \restriction n]\!]$ so that $x \restriction n \in T$. It follows that $x \in [T]$. For the converse, suppose that $x \notin K$. Since K is closed, there must be some basic interval $[\![\sigma]\!]$ containing x such that $K \cap [\![\sigma]\!] = \emptyset$. Since $x \in [\![\sigma]\!]$, this means that $\sigma = x \restriction n$ for some n. Then $x \restriction n \notin T$ and hence $x \notin [T]$. It is left to the reader to check that $[T]$ is a closed set for any tree T. $\square$

The Kleene–Brouwer order $<_{KB}$ on ω^* is defined by $\sigma \leq_{KB} \tau$ if either $\tau \sqsubseteq \sigma$ or if $\sigma(n) < \tau(n)$, where n is the least such that $\sigma(n) \neq \tau(n)$. Then the initial segments of any $x \in \mathbb{N}^\mathbb{N}$ form an infinite descending chain so that this ordering is not well founded. We can now make a general connection between finding an element of a closed set $K = [T]$ and finding a descending chain in the tree T.

Theorem 7.5.12. *For any tree $T \subseteq \omega^*$ and closed set $K = [T]$, K is empty if and only if the tree T is well founded.*

Proof. If K is non-empty, then the initial segments of any element of K form an infinite descending chain in T, hence T is not well founded. Suppose now that T is not well founded and let $\sigma_0 \succ \sigma_1 \succ \sigma_2 \succ \cdots$ be an infinite descending sequence. Then we can define an infinite path $x \in [T]$ as follows: First, observe that for each n, we must have $\sigma_{n+1}(0) \leq \sigma_n(0)$, otherwise by definition $\sigma_n \preceq \sigma_{n+1}$. This means that the values $\sigma_0(0), \sigma_1(0), \ldots$ must converge to a limit k_0, which is going to be $x(0)$, and there must be some n_0 such that $\sigma_n(0) = k_0$ for all $n \leq n_0$. Now, for $n > n_0$, it follows that $\sigma_{n+1}(1) \leq \sigma_n(1)$, and thus there will be k_1 and n_1 such that $\sigma_n(1) = k_1$ for all $n \leq n_1$. In this fashion, we can recursively define the sequences $n_0, n_1, n_2, \ldots$ and $k_0, k_1, \ldots$ and let $x = (x(0), x(1), \ldots)$. For each $n \leq n_k$, we have $(k_0, k_1, \ldots, k_n) \sqsubseteq \sigma_n$ and $\sigma_n \in T$, so that $x \upharpoonright n \in T$. It follows that $x \in [T] = K$, as desired. $\qquad \square$

Example 7.5.13. The *Stone–Čech compactification of* ω is the following space denoted by $\beta\omega$: its underlying set is the set of all ultrafilters on ω, and the topology is generated by the subbasis consisting of all sets of the form $\{u : a \in u\}$, where $a \subseteq \omega$ is an arbitrary set.

Other examples of topological spaces are obtained by applying certain operations to preexisting spaces.

Example 7.5.14. Suppose that (X, T) is a topological space and $Y \subseteq X$. The *inherited topology* $T \upharpoonright Y$ is the collection $\{A \cap Y : A \in T\}$.

In this way, we consider for example intervals $[0, 1]$ or $(0, 1) \subseteq \mathbb{R}$ with the inherited topology as topological spaces. Another example is the space of irrationals as a subspace of $[0, 1]$ or of the whole line.

Example 7.5.15. Suppose that (X_0, T_0) and (X_1, T_1) are topological spaces. The *product space* is $(X_0 \times X_1, U)$ where U is the topology on $X_0 \times X_1$ generated by the subbasis consisting of all sets of the form $O \times P$ where $O \in T_0$ and $P \in T_1$. We refer to U as the *product topology* on $X_0 \times X_1$.

In this way, we consider for example the Euclidean spaces $\mathbb{R}^2 = \mathbb{R} \times \mathbb{R}$, $\mathbb{R}^3 = \mathbb{R}^2 \times \mathbb{R}$, and in general $\mathbb{R}^n$ for any natural number $n \in \omega$ with the product topology. These spaces are pairwise non-homeomorphic — the proof of this statement was the beginning of the field of *dimension theory*.

Example 7.5.16. Suppose that I is a set and (X_i, T_i) for $i \in I$ are topological spaces. The *product space* is the pair $(\prod_i X_i, U)$, where $\prod_i X_i = \{f : Dmn(f) = I \wedge (\forall i \in I) f(i) \in X_i\}$ and U is generated by the subbasis consisting of all sets of the form $\{f \in \prod_i X_i : f(j) \in O\}$, where $j \in I$ is an index and $O \in T_j$ is an open subset of X_j.

The most notorious space obtained in this way is the *Hilbert cube* $[0, 1]^\omega$, the product of countably many copies of the interval $[0, 1]$. The Cantor space and the Baire space may both be viewed as product spaces where the sets $\{0, 1\}$ or ω are given the discrete topology.

The following notions are ubiquitous in the treatment of topological spaces:

Definition 7.5.17. Let (X, T) be a topological space. A set $D \subseteq X$ is *dense* in the space if every non-empty open set $O \in T$ contains an element of D.

Definition 7.5.18. A topological space (X, T) is *separable* if it contains a countable dense set.

The space $\mathbb{R}$ of real numbers has the rationals as a countable dense set. The Cantor space $\{0, 1\}^\omega$ has a countable dense set D consisting of those sequences x such that $\{n : x(n) = 1\}$ is finite. A similar definition provides a countable dense set for the Baire space.

A very common way of defining a topology on a space X is by way of a *metric*.

Definition 7.5.19. A *metric* on a set X is a function $d : X^2 \to \mathbb{R}$ such that

1. for every $x, y \in X$, $d(x, y) \geq 0$ and $d(x, y) = 0 \leftrightarrow x = y$;
2. for every $x, y \in X$, $d(x, y) = d(y, x)$;
3. (the *triangle inequality*) for every $x, y, z \in X$, $d(x, z) \leq d(x, y) + d(y, z)$.

A pair (X, d) where d is a metric on X is called a *metric space*.

Example 7.5.20. The *discrete metric* on any set X, assigning distance 1 between any two distinct points, is a metric. The *Euclidean metric* on $\mathbb{R}^n$ is a metric for every n. The *Manhattan metric* is a different metric on $\mathbb{R}^n$, defined by $d(x, y) = \sum_{i \in n} |x(i) - y(i)|$. The unit sphere S^2 in $\mathbb{R}^3$ can be equipped with at least two natural metrics: the metric inherited from the Euclidean metric on $\mathbb{R}^3$, or the Riemann surface metric defined by $d(x, y)$ = the length of the shorter portion of the great circle (i.e. a circle on the surface of S^2 of maximal circumference) connecting x and y.

Example 7.5.21. For the Cantor space, we let $d(x, y) = 2^{-n-1}$, where n is the least such that $x(n) \neq y(n)$.

Definition 7.5.22. If (X, d) is a metric space, $x \in X$ is a point and $\varepsilon > 0$ is a real number, the *open ball* $B(x, \varepsilon)$ is the set $\{y \in X : d(x, y) < \varepsilon\}$, and a *closed ball* $\bar{B}(x, \varepsilon)$ is the set $\{y \in X : d(x, y) \leq \varepsilon\}$. Then the *topology generated by* d on the set X is the topology generated by the open balls $B(x, \varepsilon)$ for $x \in X$ and real $\varepsilon > 0$. A topology on the set X is *metrizable* if there is a metric which generates it.

Exercises for Section 7.5

Exercise 7.5.1. Show that $\{\sigma \in \omega^* : K \cap [\![\sigma]\!] \neq \emptyset$ defined in the proof of Proposition 7.5.11 is a tree.

Exercise 7.5.2. Show that $[T]$ is a closed set for any tree T, as claimed in Proposition 7.5.11.

Exercise 7.5.3. Show that the intervals in the Cantor space form a basis. Show that the complement of any interval is also an open set.

Exercise 7.5.4. Let (X, T) be a topological space and let $B \subseteq X$ be a set. Prove that there is the inclusion-smallest closed set $C \subseteq X$ which contains B as a subset. C is referred to as the *closure* of the set B, often denoted by $\bar{B}$.

Exercise 7.5.5. Let (X, S) and (Y, T) be topological spaces. Consider the space $X \times Y$ with the product topology. Prove that the *projection function* $f : X \times Y \to X$ given by $f(x, y) = x$ is continuous.

Exercise 7.5.6. Let (X, T) be a topological space. Consider the space $X \times X$ with the product topology. Show that the function $f : X \to X \times X$ given by $f(x) = (x, x)$ is continuous.

Exercise 7.5.7. Let X_0, X_1, Y_0, Y_1 be topological spaces with their topologies and let $f_0 : X_0 \to Y_0$ and $f_1 : X_1 \to Y_1$ be continuous functions. Conclude that the function $g : X_0 \times X_1 \to Y_0 \times Y_1$ given by $g(x_0, x_1) = (f_0(x_0), f_1(x_1))$ is continuous.

Exercise 7.5.8. Let (X, S) and (Y, T) be topological spaces, and $f : X \to Y$ be a continuous function. Show that f, viewed as a subset of $X \times Y$, is a closed subset of $X \times Y$.

Exercise 7.5.9. Let X, Y, Z be topological spaces, and f, g be continuous functions from X, Y respectively to Z. Show that the set $C = \{(x, y) \in X \times Y : f(x) = g(y)\}$ is closed. Similarly, show that if X_n for $n \in \omega$ are topological spaces and $f_n : X_n \to Z$ are continuous functions then the set $C = \{u \in \prod_n X_n : \forall n, m \; f_n(u(n)) = f_m(u(m))\}$ is closed in $\prod_n X_n$.

Exercise 7.5.10. Verify that the metric defined in Example 7.5.21 for the Cantor space is in fact a metric.

Chapter 8

Models of Set Theory

Advanced set theory is concerned, in great part, with *independence results* — theorems saying that certain statements cannot be proved from the ZFC axioms. As a famous example, neither the Continuum Hypothesis nor its negation can be proved in ZFC. The main method for proving such independence results is the construction of models of set theory.

In this section, we provide a humble introduction to the construction of models of set theory. We show that certain axioms of ZFC cannot be derived from others.

In Section 8.1, we examine the finite levels V_n of the hierarchy of sets as well as the first infinite level V_ω. We give a characterization of V_ω as the set of hereditarily finite sets. We explain what it means for one of the axioms to hold in a set, as opposed to being true in the universe of all sets. We determine which of the axioms of ZFC are true in V_n and in V_ω. In particular, we show that V_ω satisfies every axiom except for the Axiom of Infinity. This demonstrates that the Axiom of Infinity is independent of the other axioms.

In the second section, we generalize to the transfinite levels V_α of the hierarchy of sets. We also generalize from the hereditarily finite sets to the hereditarily countable sets and beyond.

8.1   The Hereditarily Finite Sets

Recall the finite levels V_n of the hierarchy of sets defined in Example 4.3.15, where $V_0 = \emptyset$, $V_{n+1} = \mathcal{P}(V_n)$ for each $n \in \omega$, and $V_\omega = \bigcup_n V_n$.

The set V_ω may be viewed as a reasonable universe for doing finite mathematics. We show that it satisfies all of the axioms except for the Axiom of Infinity. One might think that, for a very large n, even V_n might suffice. However, the finite levels V_n do not satisfy many of the axioms.

We now present the characterization of V_ω as the *hereditarily finite* sets. Intuitively, a set A is hereditarily finite if A is finite, all of its elements are finite, all of the elements of the elements of A are finite, and so on.

Definition 8.1.1. A set A is hereditarily finite if its transitive closure $\mathtt{trcl}(A)$ is finite.

The following lemma is left as an exercise:

Lemma 8.1.2. *If B is hereditarily finite, then for all $A \in B$, A is hereditarily finite.*

Proposition 8.1.3. *For any set A, A is hereditarily finite if and only if $A \in V_\omega$.*

Proof. First, we recall that for any $n \in \omega$, V_n is transitive. Now, suppose that $A \in V_\omega$. Then $A \in V_{n+1}$ for some n so that $A \subseteq V_n$. But this means that $\mathtt{trcl}(A) \subseteq \mathtt{trcl}(V_n) = V_n$. Since V_n is finite, it follows that $\mathtt{trcl}(A)$ is finite, and therefore A is hereditarily finite.

Next, suppose that A is hereditarily finite, that is, $\mathtt{trcl}(A)$ is finite. Suppose by way of contradiction that $A \notin V_\omega$. Then $\{x \in \mathtt{trcl}(A) : x \notin V_\omega\}$ is non-empty and thus has an $\in$-minimal element B. B is finite, since $\mathtt{trcl}(B) \subseteq \mathtt{trcl}(A)$. By minimality, every element of B is in V_ω. Let n be the least such that each element of B is in V_n; this exists since $V_i \subseteq V_{i+1}$ for all i. Then $B \in V_{n+1}$, contradicting the assumption that $B \notin V_\omega$. $\qquad\square$

Next, we return to consideration of the axioms of ZF. The understanding is that these axioms are true in the universe V of all sets. There is one exception that we do not consider the Axiom of Choice to necessarily be true in the universe. So, we have to state when we are using this axiom.

We now examine what it means for one or more of the axioms to be true in a *set* M. For the most part, we want to consider only transitive sets M. For instance, for the Pairing Axiom, this should mean that for any two sets $x, y \in M$, the set $\{x, y\}$ belongs to M. Thus, if $0 \in M$ and M satisfies the Pairing Axiom, then $\{0\} \in M$, $\{\{0\}\} \in M$, and so on. This implies that M must be infinite.

To see why transitivity is important, consider the example $M = \{1, 2, 3\}$. This is not transitive since $0 \in 1 \in M$ but $0 \notin M$. For the two elements 1 and 2, we see that $(\forall x \in M)\, x \in 3 \iff (x = 1 \vee x = 2)$. Thus, 3 acts as the pair of 1 and 2. On the other hand, there is no set which acts as $\{3\}$ in M since no element of M contains 3. Even if a set A is not transitive, the argument above still shows that A must be infinite in order for it to satisfy the Pairing Axiom. Similar complications arise in non-transitive sets A for the Union and Power Set Axioms, in that a set can act in A as the union or power set of another set, without being the actual union or power set.

Definition 8.1.4. Let ϕ be a formula of set theory and M a transitive set. Then ϕ^M denotes the formula obtained from ϕ by restricting all its quantifiers to the set M. We say that the structure $\langle M, \varepsilon \rangle$ *satisfies* ϕ, in symbols $M \models \phi$, if ϕ^M holds.

It is important to observe that ϕ and ϕ^M are in principle different statements and that one of them may fail while the other may hold. In an important class of formulas though, the truth values of ϕ and ϕ^M coincide.

The notion of a *bounded* formula will be useful here.

Definition 8.1.5. A formula ϕ is *bounded* if all of its quantifiers are bounded, i.e. of the form $\forall x \in y$ or $\exists x \in y$.

Example 8.1.6. The following can be defined by bounded formulas: $x \subseteq y$, $x = y \cup z$, x is a binary relation on y, x is a function, and x is an inductive set.

Lemma 8.1.7. *Suppose that ϕ is a bounded formula and M is a transitive set. Then for any elements $\vec{A}$ of M, $\phi(\vec{a})$ holds if and only if $M \models \phi(\vec{a})$.*

Proof. When we check whether $(\exists x \in y)\, \phi(x, y, \dots)$ for $y \in M$, we just observe that whenever $x \in y \in M$, we have $x \in M$ as well. $\quad \square$

Now, we come to the heart of the matter. With an appropriate formalization of the proof system for first-order logic, one can show that if a sentence ϕ can be derived from a set Γ of other sentences in the language of set theory, then for every transitive set M, if $M \models \psi$ for every sentence $\psi \in \Gamma$, then $M \models \phi$. This feature is referred to as the soundness of first-order logic. Thus, to show that a sentence ϕ, one of the axioms of set theory, cannot be derived from the others, it is enough to produce a transitive set M which satisfies all axioms of ZFC with the exception of ϕ. We see in the following that V_ω satisfies all of the axioms except for the Axiom of Infinity, and this shows that the Axiom of Infinity cannot be derived from the other axioms.

Now, we examine whether the axioms are true for the finite models $M = V_n$ and for the model V_ω.

It is easy to see that the empty set V_0 itself satisfies all axioms except for the Null Set Axiom and the Axiom of Infinity. For example, $\emptyset$ satisfies the Pair Axiom vacuously, since there are no elements to make into pairs.

In this section, we begin by looking at the finite levels V_n and the first infinite level V_ω of the hierarchy of sets:

(1) **Empty Set:** For any $n > 0$, $0 \in V_n$ so that each for each $n > 0$, V_n satisfies the Empty Set Axiom, and V_ω also satisfies the Empty Set Axiom.

The following lemmas are needed for discussion of extensionality:

Lemma 8.1.8. *For any sets A, B, and C, if C is transitive and $A, B \in C$, then $A = B \iff A \cap C = B \cap C$, that is, $A = B$ if and only if, for all $x \in C$, $x \in A \iff x \in B$.*

Proof. Suppose that C is transitive and $A, B \in C$. If $A = B$, then certainly $x \in A \iff x \in B$ for any $x \in C$. Suppose now that $A \cap C = B \cap C$ and let $x \in A$. Since C is transitive, it follows that $x \in C$. Thus, by assumption $x \in B$. Similarly, $x \in B$ implies $x \in A$. Then by the Axiom of Extensionality, $A = B$. $\qquad\square$

Lemma 8.1.9. *Any transitive set M satisfies the Axiom of Extensionality.*

Proof. Suppose that $A, B \in M$ and that M satisfies $A = B$, that is, $(\forall x \in M)\, x \in A \iff x \in B$. It follows from the Lemma 8.1.8 that $A = B$. $\qquad\square$

(2) **Extensionality:** Since each V_n is transitive, and V_ω is transitive, by Proposition 4.3.26, it follows from Lemma 8.1.8 that these sets satisfy the Axiom of Extensionality.

(3) **Pairing:** For each n, V_{n+1} does not satisfy the Pairing Axiom, since $n \in V_{n+1}$, but $\{n\} \notin V_{n+1}$. However, V_ω does satisfy the Pairing Axiom.

(4) **Powerset:** V_{n+1} does not satisfy the Power Set Axiom for any n. This is left as an exercise. V_ω does satisfy this axiom. That is, given $A \in V_\omega$, we have $A \in V_{n+1}$ for some $n \in \omega$. Thus, $A \subseteq V_n$ and hence any subset of A is also a subset of V_n and therefore an element of V_{n+1}. It follows that $\mathcal{P}(A) \subseteq V_{n+1}$ and therefore $\mathcal{P}(A) \in V_{n+2}$ so that $\mathcal{P}(A) \in V_\omega$.

(5) **Union:** V_n satisfies the Union Axiom for every $n \in \omega$. To see this, first note that $\bigcup V_n \subseteq V_n$ since V_n is transitive. $V_0 = \emptyset$, which vacuously satisfies Union. $V_1 = \{0\}$ and $\bigcup 0 = 0$, so V_1 also satisfies Union. Proceeding by induction, suppose that $A \in V_{n+2}$. If $x \in y \in A$, then $y \in V_{n+1}$ and $x \in V_n$. Hence, $\bigcup A \subseteq V_n$ and thus $\bigcup A \in V_{n+1}$ and then $\bigcup A \in V_{n+2}$ as well. It is left as an exercise to show that V_ω satisfies the Union Axiom.

(6) **Infinity:** If M is transitive and $A \in M$ is inductive, then it must contain $0, 1, 2, \ldots$ for all n and therefore A must be infinite. So, no V_n can satisfy the Axiom of Infinity and also V_ω does not satisfy the axiom since all of its elements are finite.

(7) **Comprehension:** V_n satisfies Comprehension for all $n \in \omega$, as does V_ω. The proof is left as an exercise.

(8) **Replacement:** The Axiom of Replacement fails in each V_{n+1}. To see this, observe that $n - 1$ and n belong to V_{n+1}, as does the set $\{n - 1\}$. The successor function $F(x) = x \cup \{x\}$ is a class function. However, the image $F[\{n - 1\}] = \{n\}$ does not belong to V_{n+1}.

V_ω does satisfy Replacement. To see this, suppose that $A \in V_\omega$ and that F is a class function such that, for all $x \in A$, $F(x) \in V_\omega$. Let F be defined by formula $\phi(x, y)$ so that $F(x) = y \iff \phi(x, y)$. The key here is that A is a finite set. For each $x \in A$, we have $F(x) \in V_\omega$ so that $F(x) \in V_{n+1}$ for some n. Let $g(x)$ be the least n

such that $F(x) \in V_{n+1}$. Then $g[A]$ is a finite subset of ω and hence has a maximum element m so that for all $x \in A$, $F(x) \in V_{m+1}$. Then $F[A]$ is a subset of V_{m+1} and therefore is an element of V_{m+2} and hence an element of V_ω.

(9) **Axiom of Choice:** Let us consider a function-free formulation of Choice, as follows:

Proposition 8.1.10. *For any finite set A of disjoint sets, there is a set C containing exactly one element from each set in A.*

Proof. Let $A = \{x_1, \ldots, x_k\}$ be a disjoint family of non-empty sets. Since each x_k is non-empty, it follows without using the Axiom of Choice that there exist $y_i \in x_i$ for $i = 1, 2, \ldots, k$. It can be seen by induction on n that $\{y_1, \ldots, y_n\}$ is a set for each $n \leq k$. This uses the Pairing Axiom and the Union Axiom. For $n = 2$, this is immediate. Given the set $\{y_1, \ldots, y_n\}$, we can create the set $\{\{y_1, \ldots, y_n\}, \{y_{n+1}\}\}$. Then the union of this set is $\{y_1, \ldots, y_{n+1}\}$. Since each $y_i \in x_i \in A$, we have $C \subseteq \bigcup A$. $\qquad\square$

Proof of the following two propositions are left to the exercises:

Proposition 8.1.11. *For each n, V_n satisfies the above function-free Axiom of Choice.*

Proposition 8.1.12. *V_ω satisfies the standard Axiom of Choice.*

(10) **Regularity:** Given that Regularity holds in the universe of sets, let M be any set and $A \in M$ be a non-empty set. Suppose first that $A \cap M = \emptyset$. Then A is a minimal element of M by Exercise 3.8.1. Next, suppose that $A \cap M$ is not empty and let y be $\in$-minimal in $A \cap M$. We claim that y acts as a $\in$-minimal element of A in M. That is, suppose that $z \in y$, where $z \in A$ and also $z \in M$. Then $z \in A \cap M$, contradicting the assumption that y is $\in$-minimal in $A \cap M$. It follows any set M will satisfy the Axiom of Regularity.

Putting these together, we have the following:

Theorem 8.1.13. *V_ω satisfies every axiom of ZFC except for the Axiom of Infinity.*

This demonstrates that the Axiom of Infinity is independent of the other axioms. We have also shown the following:

Proposition 8.1.14. *For each $n \in \omega$, V_{n+1} satisfies every axiom except Pairing, Power Set, Infinity, and Replacement.*

We consider transfinite models of ZFC in the following section.

Exercises for Section 8.1

Exercise 8.1.1. Determine which of the following sets satisfies the Axiom of Extensionality. Recall that $0 = \emptyset$, $1 = \{0\}$, and so on:

1. $\{1, 2\}$.
2. $\{1, \{1, 2\}, \{1, 2, 3\}\}$.
3. $\{0, 1, 2, \{2\}\}$.

Exercise 8.1.2. Which Axioms of ZF are satisfied by the set $\{0, \{0\}, \{\{0\}\}, \dots\}$?

Exercise 8.1.3. Show that if B is hereditarily finite, then for all $A \in B$, A is hereditarily finite.

Exercise 8.1.4. Prove that for each n, V_n is finite.

Exercise 8.1.5. Prove each of the following:

1. For each $n \in \omega$, V_n is transitive.
2. V_ω is transitive.

Exercise 8.1.6. Show that V_4 does not satisfy the Axiom of Choice, that is, find in V_4 a set A of non-empty sets such that no function in V_3 satisfies $f(a) \in A$ for each $a \in A$.

Exercise 8.1.7. Show that for each n, V_n satisfies the function-free Axiom of Choice as in Proposition 8.1.10.

Exercise 8.1.8. Show that V_ω satisfies the Axiom of Choice. *Hint:* Prove by induction on n that for any set $A = \{x_1, \dots, x_k\} \in V_\omega$ of non-empty sets, there is a function $f = \{(x_1, y_1), \dots, (x_n, y_n)\} \in V_\omega$ with each $y_i \in x_i$.

Exercise 8.1.9. Show that no non-empty finite set can satisfy the Power Set Axiom.

Exercise 8.1.10. Give an example of a transitive set A such that $(A, \in)$ does not satisfy the Union Axiom.

Exercise 8.1.11. Show that the Empty Set Axiom follows from the Axiom of Infinity and the Axiom of Separation.

Exercise 8.1.12. Show that V_ω satisfies the Pairing Axiom.

Exercise 8.1.13. Show that V_ω satisfies the Union Axiom.

Exercise 8.1.14. Show that V_n satisfies the Axiom of Comprehension for every $n \in \omega$ and thus that V_ω satisfies it as well.

8.2 Transfinite Models

In this section, we examine the transfinite levels V_α of the hierarchy of sets, as well as the hereditarily countable sets and more generally the hereditarily $< \kappa$ sets for any cardinal κ. This reveals that the Axiom of Replacement and the Powerset Axiom are each independent of the other axioms.

The conclusions of the previous section for the finite levels of the hierarchy of sets may be applied to the transfinite successor levels except that ω belongs to any transfinite level $V_{\alpha+1}$ so that $V_{\alpha+1}$ satisfies the Axiom of Infinity.

Proposition 8.2.1. *For each infinite successor ordinal α, V_α satisfies every axiom except Pairing, Power Set, and Replacement.*

The limit levels V_λ, for $\lambda > \omega$, share most of the properties of V_ω, with two notable exceptions. Certainly, each such V_λ satisfies the Axiom of Infinity.

Thus, we have the following:

Theorem 8.2.2. *For any limit ordinal $\lambda > \omega$, V_λ satisfies every axiom of ZF except possibly Replacement.*

A situation in which V_λ does not satisfy Replacement is set forth in the following result.

Theorem 8.2.3. *For any ordinal λ such that the cofinality $\mathrm{cof}(\lambda) < \lambda$, V_λ does not satisfy the Replacement Axiom.*

Proof. Let $\kappa = cof(\lambda) < \lambda$ and let $F : \kappa \to \lambda$ be cofinal, that is, $\bigcup F[\kappa] = \lambda$. While F does not belong to V_λ, it is nevertheless a class function mapping a set $\kappa \in V_\lambda$ to V_λ. If V_λ satisfied the Replacement Axiom, then $F[\kappa]$ would belong to V_λ, and then the Union Axiom would imply that $\lambda \in V_\lambda$, a contradiction. $\square$

Note that in particular, if λ is not a cardinal, then $cof(\lambda) < \lambda$, and hence V_λ does not satisfy Replacement.

Theorem 8.2.4. *The axiom of Replacement is not provable from the other axioms of ZFC.*

Proof. We have shown that $V_{\omega+\omega}$ satisfies all axioms of ZFC except for certain instances of the axiom schema of Replacement. $\square$

We must observe that being a regular cardinal is not enough for Replacement. For example, $V_{\aleph_1}$ does not satisfy Replacement. This follows from our following result. Recall that a cardinal κ is strongly inaccessible if it is regular and if, for any $\lambda < \kappa$, $2^\kappa < \lambda$.

Theorem 8.2.5. *If κ is a cardinal and V_κ satisfies the Axiom of Replacement, then κ is strongly inaccessible.*

Proof. We know that κ must be regular by Theorem 8.2.3. Now, suppose that $\lambda < \kappa$ but $\kappa \le 2^\lambda$. This means that there is a function F mapping $\mathcal{P}(\lambda)$ onto κ. Since $\lambda < \kappa$, $\mathcal{P}(\lambda) \in V_\kappa$, so the Replacement Axiom would imply that $F[\mathcal{P}(\lambda)] = \kappa$ is an element of V_κ, a contradiction. $\square$

The converse of Theorem 8.2.5 is shown in the following, using the following lemma, which is an immediate consequence of Theorem 6.3.32. The proof is left as Exercise 8.2.8.

Lemma 8.2.6. *If κ is strongly inaccessible and $A \in V_\kappa$, then $|A| < \kappa$.*

Theorem 8.2.7. *(ZFC) V_κ satisfies the Axiom of Replacement if and only if κ is strongly inaccessible.*

Proof. It remains to show that V_κ satisfies Replacement if κ is strongly inaccessible. Let $A \in V_\kappa$ and let $F : A \to V_\kappa$ be a class function. By Lemma 8.2.6 from Chapter 6, $|A| = \lambda < \kappa$, so there is a map G from λ onto A and then $F \circ G$ maps λ to V_κ so that

$F[A] = (F \circ G)[\lambda]$. For each $\alpha < \lambda$, let $H(\alpha)$ be the least ordinal β such that $(F \circ G)(\alpha) \in V_{\beta+1}$. Then H is a class function mapping λ into κ. Since κ is inaccessible, it follows that $H[\lambda]$ is not cofinal in κ. Thus, there is some cardinal $\mu < \kappa$ such that $H(\alpha) < \mu$ for all $\alpha < \lambda$. This means that $F[A] \subseteq V_\mu$ and hence $F[A] \in V_\kappa$. $\qquad\square$

Thus, we have shown the following:

Theorem 8.2.8. *If κ is strongly inaccessible, then V_κ satisfies every axiom of ZFC.*

This explains why it is not possible to construct a strongly inaccessible cardinal within ZFC. The explanation goes beyond the scope is this book, but here is a brief explanation. Suppose there were a proof in ZFC that a strongly inaccessible cardinal κ exists. Then as we have seen, V_κ would be a model of ZFC. But this would imply that ZFC is consistent. However, Gödel's classic Incompleteness Theorem implies that no reasonably strong theory, such as ZFC, can prove its own consistency.

Since V_ω was identified with the family of hereditarily finite sets, there is another possible generalization of V_ω which is a good candidate for a model of set theory.

Definition 8.2.9. For any infinite cardinal number κ, let $H(\kappa)$ be the family of sets A such that $|\mathtt{trcl}(A)| < \kappa$. Thus, the hereditarily finite sets are the sets in $H(\omega)$. The sets in $H(\aleph_1)$ are known as the *hereditarily countable* sets; let us denote this class by HC.

HC is a set, since it can be shown that $HC \subseteq V_{\aleph_1}$ (see the exercises).

Lemma 8.2.10. *HC is a transitive set.*

Proof. Let $x \in HC$ and $y \in x$. Then $\mathtt{trcl}(y) \subseteq \mathtt{trcl}(x)$, and since the set $\mathtt{trcl}(x)$ is countable, so is $\mathtt{trcl}(y)$. This means that $y \in HC$ and so the set HC is transitive as desired. $\qquad\square$

It is left as an exercise to show that $H(\kappa)$ is transitive for every κ.

It is clear that HC does not satisfy the Powerset Axiom. However, all other axioms are satisfied.

Lemma 8.2.11. *If the set A is countable, and every element of A is hereditarily countable, then A is hereditarily countable.*

Proof. We claim that $\mathtt{trcl}(A) = \{A\} \cup \bigcup_{a \in A} \mathtt{trcl}(a)$. Since A is countable, and $\mathtt{trcl}(a)$ is countable for each $a \in A$, it follows that $\mathtt{trcl}(A)$ is countable, and hence $A \in HC$. To prove the claim, it suffice to show that $\{A\} \cup \bigcup_{a \in A} \mathtt{trcl}(a)$ is transitive. This is left as an exercise. □

Proposition 8.2.12. *HC satisfies the Axiom of Replacement.*

Proof. Let $F : HC \to HC$ be a class function and let $A \in HC$. Then $F(a) \in HC$ for every $a \in A$ and $F[A] \subseteq HC$ exists by the Axiom of Replacement. We need to show that $F[A]$ is hereditarily countable. Now, $F[A]$ is countable since A is countable, and it follows by Lemma 8.2.11 that $F[A] \in HC$. □

It is left as an exercise to show that every $H(\kappa)$ satisfies the Axiom of Replacement.

Theorem 8.2.13. *HC satisfies every axiom of ZFC except the Powerset Axiom.*

Other axioms are considered in the following exercises. Again it turns out that $H(\kappa)$ will satisfy the missing axiom, in this case, the Power Set Axiom (and thus *all* of the axioms) if and only if κ is a *strongly* inaccessible cardinal.

Thus, we have the following independence result:

Theorem 8.2.14. *The Powerset Axiom cannot be derived from the other axioms of ZFC.*

Properties of the structure $H(\kappa)$ depend on whether κ is regular or is a limit cardinal.

Theorem 8.2.15. *(AC) $H(\kappa)$ satisfies the Powerset Axiom if and only if κ is a strong limit cardinal.*

Proof. Suppose that κ is a strong limit cardinal and let $A \in H(\kappa)$. Then $|A| = \lambda$ for some cardinal $\lambda < \kappa$. Since κ is a strong limit cardinal, it follows that $|\mathcal{P}(A)| = 2^\lambda < \kappa$. For any $B \in \mathcal{P}(A)$, we have $B \subseteq A$ so that $|B| \leq |A| < \kappa$. Thus, $\mathcal{P}(A) \in H(\kappa)$.

The converse direction is left as an exercise. □

The Union Axiom also comes into play for the sets $H(\kappa)$. For example, let $\kappa = \aleph_\omega$. Then the set $A = \{\aleph_n : n \in \omega\}$ belongs to

$H(\kappa)$, but $\bigcup A = \aleph_\omega$ does not. Hence, $H(\aleph_\omega)$ does not satisfy the Union Axiom. Similarly, $H(\beth_\omega)$ does not satisfy the Union Axiom.

Theorem 8.2.16. *The Union Axiom cannot be derived from the other axioms of ZFC.*

Proof. The model $H(\beth_\omega)$ does not satisfy the Union Axiom. It satisfies the Powerset axiom since $\beth_\omega$ is a strong limit cardinal, and it satisfies the other axioms since every $H(\kappa)$ satisfies them (for $\kappa > \aleph_0$). □

Here is the general result for the Union Axiom.

Theorem 8.2.17. *(AC) $H(\kappa)$ satisfies the Union Axiom if and only if κ is a regular cardinal.*

Proof. Suppose that κ is regular and let $A \in H(\kappa)$. For any $B \in A$, $B \in H(\kappa)$ as well so that $\bigcup A \subseteq H(\kappa)$, and each $B \in A$ has cardinality $< \kappa$. It now follows from Theorem 6.3.28 that $|\bigcup A| < \kappa$. Thus, $\bigcup A \in H(\kappa)$ as desired. The converse direction is left as an exercise. □

Putting this all together, we have the following:

Theorem 8.2.18. *If κ is a strongly inaccessible cardinal, then $H(\kappa)$ satisfies every axiom of ZFC.*

One might speculate that $H(\kappa)$ equals V_κ for any strongly inaccessible cardinal κ. Showing this is left as an exercise.

Exercises for Section 8.2

Exercise 8.2.1. Show that for any limit ordinal λ, V_λ satisfies the Pairing Axiom.

Exercise 8.2.2. Show that V_α satisfies the Axiom of Comprehension for all ordinals α.

Exercise 8.2.3. Show that $V_{\alpha+1}$ does not satisfy the Power Set Axiom for any α.

Exercise 8.2.4. Show that for each n, V_n satisfies the function-free Axiom of Choice as in Proposition 8.1.10.

Exercise 8.2.5. Show that V_ω satisfies the Axiom of Choice. *Hint*: Prove by induction on n that for any set $A = \{x_1, \ldots, x_k\} \in V_\omega$ of non-empty sets, there is a function $f = \{(x_1, y_1), \ldots, (x_n, y_n)\} \in V_\omega$ with each $y_i \in x_i$.

Exercise 8.2.6. Show that for any countable ordinal $\alpha > \omega$, V_α does not satisfy the Axiom of Replacement. *Hint*: Use the fact that there is a bijection between α and ω.

Exercise 8.2.7. Show that if κ is a successor cardinal, then V_κ cannot satisfy Replacement.

Exercise 8.2.8. Show that if κ is strongly inaccessible and $A \in V_\kappa$, then $|A| < \kappa$.

Exercise 8.2.9. Show that $HC \subseteq V_{\aleph_1}$. *Hint*: Use set induction.

Exercise 8.2.10. Show that HC satisfies the Pairing Axiom.

Exercise 8.2.11. Show that HC satisfies the Union Axiom.

Exercise 8.2.12. Show that for any cardinal κ, $H(\kappa)$ is transitive.

Exercise 8.2.13. Show that for any cardinal κ, $H(\kappa)$ satisfies the Axiom of Replacement.

Exercise 8.2.14. Show that if $H(\kappa)$ satisfies the Powerset Axiom, then κ is a strong limit cardinal.

Exercise 8.2.15. Show that if $H(\kappa)$ satisfies the Union Axiom, then κ is a regular cardinal.

Exercise 8.2.16. Show that $\{A\} \cup \bigcup_{a \in A} \mathtt{trcl}(a)$ is transitive.

Exercise 8.2.17. Show that no nonempty finite set can satisfy the Pairing Axiom.

Exercise 8.2.18. Show that for any limit ordinal λ, V_λ satisfies the Power Set Axiom.

Exercise 8.2.19. Show that $H(\kappa)$ equals V_κ for any strongly inaccessible cardinal κ.

Chapter 9

Ramsey Theory

In this chapter, we present the basics of Ramsey theory, an area with many applications throughout mathematics. It exploits a simple, yet profound, idea: In every large enough random structure, traces of order can be found.

9.1 Finite Patterns

We start the exposition of finite Ramsey theorems with the simplest case of them all.

Example 9.1.1. In any party of six people, there are three who know each other, or there are three who do not know each other.

Proof. Let P denote the set of people in the party. Pick a person $p \in P$ at random. One of the sets $a = \{q \in P : p \text{ and } q \text{ know each other}\}$ or $b = \{q \in P : p \text{ and } q \text{ do not know each other}\}$ has size at least three; for definiteness, assume it is the former. If there are people $q, r \in a$ who know each other, then $\{p, q, r\}$ is the required set of people who know each other; otherwise, the set a is a set of at least three people who do not know each other. □

It turns out that a similar pattern persists for arbitrarily large finite numbers.

Theorem 9.1.2. *For any numbers $m, n \geq 2$ there is a number $p \in \omega$ such that in any party of p many people, there are either m many who know each other, or n many who do not know each other.*

Proof. Let $R(m, n)$ be the smallest number which works, if it exists. It is clear that $R(2, k) = R(k, 2) = k$: In any party of k many people, either two of them know each other or none of them know each other. For other choices of m, n, we have the following recursive formula:

Claim 9.1.3. *Suppose that $m, n \geq 3$. Then $R(m, n) \leq R(m, n-1) + R(m-1, n)$.*

Proof. Consider a party of at least $R(m, n-1) + R(m-1, n)$ many people. We must find either m many who know each other or n many who do not know each other. Pick one person in the party at random, say Sara. The rest of the people in the party divide into two groups: those who know Sara and those who do not. By the assumption on the size of the party, either the former group has at least $R(m-1, n)$ many elements or the latter group has at least $R(m, n-1)$ many elements. Suppose that the former case occurs; The latter is symmetric. In the group of people who know Sara, there must now be either n many elements who do not know each other — in which case we are done, or $m-1$ many elements who do know each other — and together with Sara this is a group of m people who know each other and we are done again. $\square$

It is now easy to argue by induction on $m + n$ that the number $R(m, n)$ exists using the claim at each stage of induction. $\square$

The numbers $R(m, n)$ (the *Ramsey numbers*) defined in the course of the proof of the previous theorem are quite difficult to evaluate already for low values of m, n. The case $R(4, 3)$ is relegated to the exercises; however, the exact values of $R(5, 5)$ or $R(6, 6)$ were unknown in 2018 even though very many people tried their luck with them. Since the proof of Theorem 9.1.2 provides an obvious upper bound on the value of $R(m, n)$, the difficulty resides only in the fact that a brute force consideration of all possible constellations takes too many computational resources.

Ramsey theory is the branch of mathematics dealing with far-reaching generalizations of Theorem 9.1.2 for finite and even infinite sets. It will be useful to establish parlance and notation to facilitate the rather long expressions.

Definition 9.1.4.

1. Let a be a set and $n \in \omega$ be a number. $[a]^n$ denotes the set of all subsets of a of cardinality n.
2. Let a be a set and $n, m \in \omega$ be numbers. A function $f \colon [a]^n \to m$ is called a *coloring* of $[a]^n$ by m colors.
3. Let $f \colon [a]^n \to m$ be a coloring. A set $b \subseteq a$ is called *homogeneous* for f if f is constant on $[b]^n$. The constant value is called the *homogeneous color*.

We observe that a coloring $f \colon [a]^n \to m$ may be viewed as a partition of $[a]^n$ into m sets.

The motivating question of Ramsey theory is as follows: Given a large set a and a coloring, can one find a (smaller, but still large) set $b \subseteq a$ that is homogeneous for the coloring? To formulate the answer succinctly, we use the arrow notation. While initially confusing, it was introduced in 1956 by Paul Erdős, one of the greatest mathematicians of the 20th century, and it has persisted since.

Definition 9.1.5. Let κ, λ, n, m be cardinal numbers. The notation $\kappa \to (\lambda)^n_m$ denotes the following statement: For every set a of cardinality at least κ and every coloring $f \colon [a]^n \to m$, there is a homogeneous set $b \subseteq a$ of cardinality at least λ.

In particular, the initial example shows that $6 \to (3)^2_2$ holds. In general, the cardinals κ, λ, n, m may be finite or infinite. Already the finite case is quite interesting. The following theorem is proved in the following section.

Theorem 9.1.6. *For all natural numbers $n, m, r \in \omega$, there is a natural number $s \in \omega$ such that $s \to (r)^n_m$.*

Many applications of the finite Ramsey-type theorems deal with configurations of points in the plane or higher-dimensional Euclidean spaces; the following is the first result of this form.

Corollary 9.1.7 (Erdős–Szekeres). *For every number $r \in \omega$, there is a number $s \in \omega$ such that among any s many points in the plane, no three of which are colinear, there are vertices of a convex r-gon.*

Proof. We start with an elementary observation.

Claim 9.1.8. *Among any five points in the plane, no three of which are colinear, there are vertices of a convex 4-gon.*

Proof. The argument proceeds essentially by considering all possible configurations. Let $\{x_i \colon i \in 5\}$ be the points in the plane. Consider the smallest convex polygon containing all of them. If it is a 4-gon or even a 5-gon, then its vertices are as required in the claim. Consider the case that this polygon is just a triangle; without loss, its vertices are x_0, x_1 and x_2. Both x_3, x_4 lie inside the triangle $x_0 x_1 x_2$. The lines $x_0 x_3$, $x_1 x_3$, and $x_2 x_3$ divide the triangle into six smaller triangular pieces, and the vertex x_4 must belong to one of them. If this piece is bounded by the lines $x_i x_3$ and $x_j x_3$, then the points x_i, x_j, x_3, x_4 form a convex 4-gon. □

The following is also needed and may be proved by induction on r.

Claim 9.1.9. *Let $r \geq 4$ and let b be any set of r points in the plane, no three of which are colinear. If any 4 of these points are the vertices of a convex 4-gon, then the points are the vertices of a convex r-gon.*

To conclude the proof of the theorem, without loss assume that $r \geq 5$. By the finitary Ramsey theorem (Theorem 9.1.6), there is a number s such that $s \to (r)^4_2$; we claim that s must work. Indeed, if $\{x_i \colon i \in s\}$ are pairwise distinct points in the plane no three of which are colinear, let $f \colon [s]^4 \to 2$ be the coloring defined by $f(c) = 0$ if the points x_i for $i \in c$ form a convex 4-gon, and let $f(c) = 1$ otherwise. The choice of the number s guarantees an existence of a homogeneous set $b \subseteq s$ of size r. The first claim shows that the homogeneous color cannot be 1, so it must be 0 and then the second claim shows that the points $\{x_i \colon i \in b\}$ are vertices of a convex r-gon. □

The question of the existence of homogeneous sets becomes much more difficult if one requires the homogeneous set to carry additional structure. This is a demanding area of research. We state several interesting theorems without proofs.

Fact 9.1.10 (Schur's Theorem). For every number $m \in \omega$ there is a number $s \in \omega$ such that for every coloring $f \colon s \to m$ there are numbers $x, y, z \in s$ such that $x + y = z$ and $f(x) = f(y) = f(z)$.

Recall that an arithmetic progression on natural numbers of length n is a sequence of the form $a, a + b, a + 2b, \ldots, a + (n-1)b$ for some $a, b \in \omega$.

Fact 9.1.11 (van der Waerden's Theorem). For numbers $n, m \in \omega$ there is a number $s \in \omega$ such that for every coloring $f: s \to m$ there is a homogeneous arithmetic progression of length n contained in s.

Fact 9.1.12 (Szemerédi's Theorem). For each number $n \in \omega$ and positive real number $\epsilon > 0$ there is a number $s \in \omega$ such that every set $a \subseteq s$ such that $|a| > \epsilon \cdot s$ contains an arithmetic progression of length n.

In all cases, the numbers s whose existence is claimed grow very fast depending on the input, and finding the smallest s possible is very demanding. Szemerédi's theorem belongs to the most elaborated mathematical results of the second half of 20th century.

Exercises for Section 9.1

Exercise 9.1.1. Show that $R(4, 3) = 9$. *Hint*: To show that $R(4, 3) > 8$, consider the party of people enumerated by numbers $0, 1, \ldots, 7$ in which numbers k, l know each other if $k - l = 2$ or 3 or 5 or 6 modulo 8.

Exercise 9.1.2. Calculate the Schur number for two and three colors.

Exercise 9.1.3. Calculate the van der Waerden number for an arithmetic sequence of length three and two colors.

Exercise 9.1.4. Let $m, n \in \omega$ be numbers. Show that if the points on the circle are colored by m many colors, there is a real number $\epsilon > 0$ and a homogeneous sequence of points on the circle of length n in which any two successive points have distance ϵ.

Exercise 9.1.5. Prove that for $n, m \in \omega$, there is a number s such that every increasing sequence $\{a_i : i \in n\}$ of natural numbers starting with 0 and such that $a_{i+1} - a_i < m$ for all i contains an arithmetic progression of length 5.

Exercise 9.1.6. Prove that there is a number s such that if $s = a_0 \cup a_1$ then either a_0 contains five consecutive natural numbers or a_1 contains an arithmetic progression of length 5.

9.2   Countably Infinite Patterns

The original Ramsey theorem was proved for the case of countably infinite homogeneous sets. Frank Ramsey was a British mathematician, philosopher, and economist and proved the theorem in 1928.

Theorem 9.2.1 (Ramsey). *If n, m are natural numbers, then $\omega \to (\omega)^n_m$ holds.*

Proof.   We proceed by induction on n. In the induction process, the following simple notion is used: If $f \colon [\omega]^{n+1} \to m$ is a coloring, a set $b \subseteq \omega$ is called *end-homogeneous* for f if the value $f(c)$ for a subset $c \subseteq b$ of size $n + 1$ does not depend on the largest element of c.

Claim 9.2.2. *For every $n, m \in \omega$ and every coloring $f \colon [\omega]^{n+1} \to m$, there is an infinite end-homogeneous set for f.*

Proof.   By induction on $j \in \omega$, we find natural numbers $i_j \in \omega$ and infinite sets $b_j \subseteq \omega$ so that

- $i_0 \in i_1 \in i_2 \in \ldots$ and $b_0 \supseteq b_1 \supseteq \ldots$;
- $i_m \in b_m$;
- whenever $c \subseteq \{i_k \colon k \in j\}$ is a set of size n, the value $f(c \cup \{t\})$ is the same for all $t \in b_j$.

To begin, let $i_0 = 0$ and $b_0 = \omega$. For the induction step, suppose that the numbers i_k for $k \leq j$ and the set b_j have been found. Write $d = \{i_k \colon k \leq j\}$. For each number $t \in b_j$, let $g_t \colon [d]^n \to m$ be the coloring defined by $g_t(c) = f(c \cup \{t\})$. Since there are only finitely many colorings from $[d]^n$ to m available, one of them has to repeat infinitely many times: There must be a coloring $g \colon [d]^n \to m$ and an infinite set $b_{j+1} \subseteq b_j$ such that $g_t = g$ for all $t \in b_{j+1}$. Let i_{j+1} be the minimal element of b_{j+1} larger than i_j. This concludes the induction step and the proof.   □

Now, back to the proof of the theorem. The statement is obviously true for $n = 1$–every partition of ω into finitely many pieces must have an infinite piece. For the induction step, suppose that the statement has been verified for some $n \in \omega$; we must check it for $n + 1$. To this end, let $f \colon [\omega]^{n+1} \to m$ be a coloring and work to find an infinite homogeneous set for it. Let $b \subseteq \omega$ be an infinite end-homogeneous set guaranteed by the claim. Let $g \colon [b]^n \to m$ be the function defined by $g(c) =$ the unique value of $f(c \cup \{j\})$ where $j \in a$ is a number larger than $\max(c)$. The function g is well-defined since the set b is end-homogeneous. By the induction hypothesis, the coloring g has an infinite homogeneous set. This set is easily seen to be homogeneous for f as well. This completes the induction step and the proof. □

The Ramsey theorems in the infinite realm almost always have "miniaturizations": purely finite little brothers (sisters?). They can be proved in different ways; we choose an argument which makes use of a non-principal ultrafilter on natural numbers.

Theorem 9.2.3. *For all natural numbers $n, m, r \in \omega$, there is a natural number $s \in \omega$ such that $s \to (r)^n_m$.*

Proof. Suppose toward contradiction that the conclusion fails for some n, m, r, and for each number $s \in \omega$ choose a coloring $f_s \colon [s]^n \to m$ which has no homogeneous set of size r. Let U be a non-principal ultrafilter on ω. Consider the U-*average* of the colorings f_s for $s \in \omega$. This is a partition $f \colon [\omega]^n \to m$ defined by $f(c) = i$ if the set $\{s \in \omega \colon f_s(c) = i\}$ belongs to U.

Claim 9.2.4. *The partition f is well defined.*

Proof. For each set $c \in [\omega]^n$, the natural numbers divide into $m+1$ many sets: The set $b = \{s \in \omega \colon s \leq \max(c)\}$ and the sets $b_i = \{s \in \omega \colon s > \max(c) \text{ and } f_s(c) = i\}$ for $i \in m$. Since U is an ultrafilter, exactly one of these pieces must belong to U. Since U is non-principal, it cannot be the case that $b \in U$ since b is finite. If $b_i \in U$ for some $i \in m$, then $f(c) = i$. □

Now, we use the infinitary Ramsey theorem to find an infinite homogeneous set $b \subseteq \omega$ for f with homogeneous color $i \in m$. Let $d \subseteq b$ be any subset of size r. For each set $c \in [d]^n$, the set $e_c = \{s \in \omega \colon f_s(c) = i\}$ belongs to the ultrafilter U. Thus, the set $\bigcap \{e_c \colon c \in [d]^n\}$ is nonempty, containing some number $s \in \omega$. But then, d is a homogeneous set for the partition f_s in color i, contradicting the choice of the coloring f_s. □

An interesting variation on Ramsey theorems are *canonical Ramsey theorems*. Instead of colorings, they homogenize equivalence relations on n-tuples. We provide the simplest case with a proof.

Definition 9.2.5. $E_{\min}$ is the equivalence relation on $[\omega]^2$ connecting two pairs of natural numbers if they have the same minimum. Similarly, $E_{\max}$ is the equivalence relation on $[\omega]^2$ connecting two pairs of natural numbers if they have the same maximum.

Theorem 9.2.6 (Erdős–Rado). *Let E be an equivalence relation on $[\omega]^2$. There is an infinite set $a \subseteq \omega$ such that either any two pairs in $[a]^2$ are E-related, or no two distinct pairs in $[a]^2$ are E-related, or $E \upharpoonright [a]^2 = E_{\min} \upharpoonright [a]^2$, or $E \upharpoonright [a]^2 = E_{\max} \upharpoonright [a]^2$.*

Proof. The argument consists of two successive applications of the Ramsey theorem. In the first, we analyze the behavior of the equivalence relation E on disjoint pairs. For a quadruple $n < m < p < r$, put $g(n, m, p, r)$ equal to

- 0 if $\{n, m\}$ is E-related to $\{p, r\}$;
- 1 if the first item fails and $\{n, r\}$ is E-related to $\{m, p\}$;
- 2 if the first two items fail and $\{n, p\}$ is E-related to $\{m, r\}$;
- 3 if the first three items fail.

The Ramsey theorem yields an infinite set $a \subseteq \omega$ homogeneous for g. There is now a discussion of the different possibilities for the homogeneous color. If the homogeneous color is 0, then any two pairs in $[a]^2$ are E-related because they are (by the homogeneity of a) E-related to all pairs in $[a]^2$ with a sufficiently large minimum and E is a transitive relation. If the homogeneous color is 1, then let $b = a \setminus \{\min(a)\}$ and argue that any two pairs in $[b]^2$ must be E-related since they are E-related to a pair $\{\min(a), n\}$ for every large enough number $n \in a$, and E is a transitive relation. The homogeneous color 2 is impossible: Looking at the first six numbers $n < m < p < r < s < t$ of a, the assumption of homogeneity would tell us that the pair $\{m, s\}$ is E-related to both $\{n, p\}$ and $\{r, t\}$; by the transitivity of E, it should be the case that $\{n, p\} \; E \; \{r, t\}$ and so $g(n, m, r, t) = 0$, contradicting the assumed homogeneity of the set a in color 2. Finally, if the homogeneous color is 3, then no two disjoint pairs in $[a]^2$ are E-related.

In the first two cases, we have proved the theorem, since the first disjunct of its conclusion has been confirmed. We thus consider the last case and analyze the behavior of the equivalence E on pairs which are not disjoint. Color triples $n < m < p$ of natural numbers according to the shape of the equivalence relation E on the triple. Put $f(n, m, p)$ equal to

- 0 if no two distinct pairs from $\{m, n, p\}$ are E-related;
- 1 if $\{n, m\}$ and $\{m, p\}$ are E-related;
- 2 if only $\{n, m\}$ and $\{n, p\}$ are E-related;
- 3 if only $\{n, p\}$ and $\{m, p\}$ are E-related.

One can verify that there are no other options for the shape of E on the triple $n < m < p$. Theorem 9.2.1 gives an infinite homogeneous set $c \subseteq a$. The proof is concluded by the discussion of all possible cases for the homogeneous color. If the homogeneous color is 0, then no two distinct pairs in $[c]^2$ are E-related. Homogeneous color 1 is impossible since then if $n < m < p < r$ are the first four numbers in c, then $\{n, m\}$ E $\{m, p\}$ E $\{p, r\}$ and, by the transitivity of E, $\{n, m\}$ E $\{p, r\}$. This contradicts the assumption that the set a is homogeneous for g in color 3. If the homogeneous color is 2, then any two pairs in c are equivalent just in case they have the same minimum. If the homogeneous color is 3, then any two pairs in c are equivalent just in case they have the same maximum. In all cases, we verified one of the conjuncts of the conclusion of the theorem. $\square$

Exercises for Section 9.2

Exercise 9.2.1. Show that every infinite sequence of pairwise distinct real numbers contains an infinite increasing subsequence or an infinite decreasing subsequence.

Exercise 9.2.2. Show that every infinite sequence of pairwise distinct ordinals contains an infinite increasing subsequence.

Exercise 9.2.3. Prove that if a is an infinite set of points in $\mathbb{R}^2$, then it contains in infinite subset $b \subseteq a$ such that no two distinct pairs in b have the same distance between their points.

9.3 Uncountable Patterns

It is much more difficult to get uncountable homogeneous sets for arbitrary partitions. In this section, we include two classical negative results with proofs.

Theorem 9.3.1 (Sierpiński). *Let κ be the cardinality of the continuum. Then $\kappa \nrightarrow (\aleph_1)_2^2$.*

Proof. Let $\langle r_\alpha : \alpha \in \kappa \rangle$ be an enumeration of all real numbers by ordinals in κ. Define a coloring $f : [\kappa]^2 \to 2$ by $f(\alpha, \beta) = 0$ if $\alpha \in \beta \leftrightarrow r_\alpha < r_\beta$ as real numbers. That is, the coloring considers the ordinal ordering on κ and the ordering induced by the real numbers and compares them. We claim that there is no uncountable homogeneous set for the coloring f. Suppose that $A \subseteq \kappa$ is such a homogeneous set; we will find an injection of A into the rational numbers, proving that A is countable. Depending on the homogeneous color, the reals $\langle r_\alpha : \alpha \in A \rangle$ form an increasing sequence, or a decreasing sequence. In either case, for each ordinal $\alpha \in A$ (except possibly the largest one, if it exists), find the smallest ordinal $\beta \in A$ larger than α and find some rational number $g(\alpha)$ which is strictly between the real numbers r_α and r_β. The function $g : A \to \mathbb{Q}$ is an injection as desired. $\square$

Increasing the value of κ much higher than the continuum, it is possible to get the statement $\kappa \to (\aleph_1)_2^2$. The following question concerns a natural generalization of the Ramsey theorem to the uncountable realm:

Question 9.3.2 (Tarski). Is there a cardinal $\kappa > \omega$ such that $\kappa \to (\kappa)_2^2$?

Cardinals with this partition property are called *weakly compact* and must stand very high in the cardinality hierarchy. The question is unresolvable in ZFC.

The second result of this section shows that if the exponent is infinite, then no increase in the size of the set serving as the basis of the coloring will help get a large homogeneous set.

Theorem 9.3.3 (Erdős). *For every cardinal κ, $\kappa \nrightarrow (\omega)_2^{\aleph_0}$.*

Proof. Fix the cardinal κ. Let E be an equivalence relation on $\mathcal{P}(\kappa)$ connecting sets b, c if the symmetric difference $b \bigtriangleup c$ is finite.

Use the Axiom of Choice to find a transversal $A \subseteq \mathcal{P}(\kappa)$ for the equivalence relation E (i.e. a set consisting of one element from each E-equivalence class). Let $f \colon [\kappa]^{\aleph_0} \to 2$ be the coloring defined by $f(b) = 0$ just in case the cardinality of $a \bigtriangleup b$ is even, where a is the unique element of A which is E-equivalent to b. We show that no infinite set can be homogeneous for the coloring f.

Suppose that $B \subseteq \kappa$ is such a homogeneous infinite set, and pick two sets $b, c \subseteq B$ which are infinite, countable, and such that $|b \bigtriangleup c| = 1$. Then b, c are E-related; let $a \in A$ be the unique point in the transversal which is E-related to both. Then $|a \bigtriangleup b|$ is even just in case $|a \bigtriangleup c|$ is odd, since the two symmetric differences differ exactly in the one point which makes distinction between b and c. It follows that $f(b) \neq f(c)$ and the set B is not homogeneous. $\quad\square$

Bibliography

[1] P. Aczel. *Non-well-founded sets*. CSLI Lecture Notes 14. Stanford University, Stanford, 1988.

[2] G. Cantor. Ueber unendliche, lineare punktmannichfaltigkeiten. *Math. Ann.*, 21:545–591, 1883.

[3] D. Cenzer, J. Larson, C. Porter and J. Zapletal. *Foundations of Mathematics*. World Scientific, 2019.

[4] A. Fraenkel. *Abstract Set Theory*. Studies in Logic and the Foundations of Mathematics. North-Holland, 1953.

[5] C. Kuratowski. Une méthode d'élimination des nombres transfinis des raisonnements mathématiques. *Fund. Math.*, 3:76–108, 1922.

[6] D. Anthony Martin. A purely inductive proof of Borel determinacy. In A. Nerode and R. A. Shore, editors, *Recursion Theory*, Proceedings of Symposia in Pure Mathematics, No. 42, pp. 303–308. American Mathematical Society, Providence, 1985.

[7] J. Mycielski and H. Steinhaus. A mathematical axiom contradicting the axiom of choice. *Bull. Acad. Polonaise Sci. Ser. Sci. Math., Astron. Phys.*, 10:1–3, 1962.

[8] W. V. Quine. *New Foundations for Mathematical Logic*, pp. 80–101. Harvard University Press, 1980.

[9] B. Russel and A. Whitehead. *Principia Mathematica*. Cambridge University Press, Cambridge, 1910.

[10] M. Souslin. Problème 3. *Fund. Math.*, 1:223, 1920.

[11] J. von Neumann. über die definition durch transfinite induktion und verwandte fragen der allgemeinen mengenlehre. *Math. Ann.*, 99: 373–391, 1928.

[12] E. Zermelo. Beweis, dass jede menge wohlgeordnet werden kann. *Math. Ann.*, 59:514–516, 1904.

Index

www.ingramcontent.com/pod-product-compliance
Lightning Source LLC
Chambersburg PA
CBHW050641190326
41458CB00008B/2366